To get extra value from this book for no additional cost, go to:

http://www.thomson.com/wadsworth/shuman

thomson.com is the World Wide Web site for Wadsworth/ITP and is your direct source to dozens of on-line resources. *thomson.com* helps you find out about supplements, experiment with demonstration software, search for a job, and send e-mail to many of our authors. You can even preview new publications and exciting new technologies.

thomson.com: *It's where you'll find us in the future.*

Multimedia in Action

James E. Shuman

BELLEVUE COMMUNITY COLLEGE

An Imprint of Wadsworth Publishing Company
I(T)P® An International Thomson Publishing Company

Belmont, CA ■ Albany, NY ■ Bonn ■ Boston ■ Cincinnati ■ Detroit ■ Johannesburg ■ London ■ Madrid
Melbourne ■ Mexico City ■ New York ■ Paris ■ Singapore ■ Tokyo ■ Toronto ■ Washington

DEDICATION

To my wife, Barbara, and our sons,
Brandon and Jeremy

New Media Publisher: Kathy Shields
Assistant Editor: Shannon McArdle
Marketing Manager: David Garrison
Project Manager: Gary Palmatier, Ideas to Images
Design and Art Direction: Gary Palmatier
Illustration: Robaire Ream, Ideas to Images
Composition: Ideas to Images
Copy Editor: Elizabeth von Radics
Print Buyer: Karen Hunt
Permissions Editor: Jeanne Bosschart
Cover Photograph: Warren Bolster
Cover Design and Realization: Gary Palmatier
Printer: World Color Book Services, Taunton

COPYRIGHT © 1998 by Wadsworth Publishing Company

A Division of International Thomson Publishing Inc.

I(T)P® The ITP logo is a registered trademark under license.

Printed in the United States of America

1 2 3 4 5 6 7 8 9 10

For more information, contact Wadsworth Publishing Company, 10 Davis Drive, Belmont, CA 94002, or electronically at http://www.thomson.com/wadsworth.html

International Thomson Publishing Europe
Berkshire House 168-173
High Holborn
London, WC1V 7AA, England

Thomas Nelson Australia
102 Dodds Street
South Melbourne 3205
Victoria, Australia

Nelson Canada
1120 Birchmount Road
Scarborough, Ontario
Canada M1K 5G4

International Thomson Publishing GmbH
Königswinterer Strasse 418
53227 Bonn, Germany

International Thomson Editores
Campos Eliseos 385, Piso 7
Col. Polanco
11560 México D.F. México

International Thomson Publishing Asia
221 Henderson Road
#05-10 Henderson Building
Singapore 0315

International Thomson Publishing Japan
Hirakawacho Kyowa Building, 3F
2-2-1 Hirakawacho
Chiyoda-ku, Tokyo 102, Japan

International Thomson Publishing Southern Africa
Building 18, Constantia Park
240 Old Pretoria Road
Halfway House, 1685 South Africa

Library of Congress Cataloging-in-Publication Data
Shuman, James E.
   Multimedia in action / James E. Shuman.
     p.   cm.
   Includes index.
   ISBN 0-534-51370-0
   1. Multimedia systems. I. Title.
QA76.575.S566   1997
006.6—dc21   97-9298

All rights reserved. No part of this work covered by the copyright hereon may be reproduced or used in any form or by any means—graphic, electronic, or mechanical, including photocopying, recording, taping, or information storage and retrieval systems—without the written permission of the publisher.

CONTENTS

Preface   ix

PART I: Introduction to Multimedia   1

1 What Is Multimedia?   3

What Is Multimedia?   4
- A Computer-based Medium   5
- Interactivity: The Key Component   5
- The Elements of Multimedia   8

The Growth of Multimedia   8
- Growth from a Marketing Standpoint   9
- Growth from a User Standpoint   10

Examples of Multimedia   11

The Major Categories of Multimedia Titles   12
- Entertainment   12
- Education   13
- Corporate Communications   14
- Reference   15

Other Categories of Multimedia   15

Delivering Multimedia   16
- Compact Disc   17
- Kiosk   17
- Online   18

Inappropriate Use of Multimedia   19
- Text-Intensive Content   19
- Linear Content   20

Some Cost-effective Alternatives   20

KEY TERMS   21
REVIEW QUESTIONS   21
PROJECTS   22

PART II: Multimedia Tools   25

2 Hardware Components of a Multimedia System   27

The Multimedia Personal Computer   29

The Playback System   31
- Processor   31
- Memory   31
- Monitor and Video Card   32
- Audio Card   35
- CD-ROM Drive   35
- Upgrade Kits and Bundled Systems   36

The Development System   37
- Processor   38
- Memory   38
- Video Capture Card   38
- Monitor   38
- Peripherals   38

KEY TERMS   41
REVIEW QUESTIONS   42
PROJECTS   42

3 Multimedia Elements: Text and Graphics   45

Working with Text   46
- Be Concise   46
- Use Appropriate Fonts   46
- Make It Readable   48
- Consider Type Styles and Colors   50
- Use Restraint and Be Consistent   50

Accommodating Text-Intensive Titles 51
  Hyperlinking 52
  Pop-up Messages, Scroll Bars, and Drop-down Boxes 52
Software for Creating and Editing Text 53
Working with Graphics 54
  Graphic Image Quality 55
  Image Size, Color Depth, and File Size 55
Software for Creating and Editing Graphics 57
Features of Graphics Programs 59
Sources of Graphic Images 62
  KEY TERMS 66
  REVIEW QUESTIONS 66
  PROJECTS 67

4 Multimedia Elements: Sound, Animation, and Video 69

Sound 70
  Sampling 71
MIDI 74
Animation 74
  2-D Animation 75
  3-D Animation 77
  Animation Special Effects 79
Virtual Reality 81
Video 82
  Digitizing the Video Signal 82
  File Size Considerations 83
  Video Compression 85
  Software for Capturing and Editing Video 88
  KEY TERMS 89
  REVIEW QUESTIONS 90
  PROJECTS 91

5 Multimedia Authoring Programs 93

Multimedia Presentations 94
Stand-alone Applications 95
How Authoring Systems Work 96
  Electronic Slide Shows 96
  The Card Stack and Book Metaphors 98
  Icon-based Authoring Programs 101
  Time-based Authoring Programs 102
Programming Languages 104
  Scripting 104
  Asymetrix ToolBook's OpenScript 105
  Macromedia Director's Lingo 107
  Custom Codes and Third-party Developers 108
  Web-based Authoring Tools 109
  KEY TERMS 110
  REVIEW QUESTIONS 110
  PROJECTS 111

PART III Developing Multimedia 113

6 Developing Multimedia Titles 115

Steps in Developing Interactive Multimedia 116
The Planning Phase 117
  Step 1: Developing the Concept 117
  Step 2: Stating the Purpose 119
  Step 3: Identifying the Target Audience 120
  Step 4: Determining the Treatment 121
  Step 5: Developing the Specifications 124
  Step 6: Storyboard and Navigation 127
The Creating Phase 131
  Step 7: Developing the Content 131
  Step 8: Authoring the Title 132

The Testing Phase 133

   Step 9: Testing the Title 133

      KEY TERMS 135
      REVIEW QUESTIONS 136
      PROJECTS 137

7 Designing for Multimedia 139

Creating the Appearance— Basic Design Principles 140

   Balance 140

   Unity 144

   Movement 146

Designing for Interactivity 147

   Audience 147

   Type of Title 148

   Content 150

Guidelines for Interactive Design 154

   Make It Simple, Easy to Understand, and Easy to Use 154

   Build in Consistency 155

   Use Design Templates 156

   Provide Feedback 157

   Provide Choices and Escapes 157

      KEY TERMS 158
      REVIEW QUESTIONS 158
      PROJECTS 159

8 Managing Multimedia Projects 161

Management Issues of Multimedia Development 162

   Who Will Manage the Development Process? 162

The Management Process and Multimedia Projects 165

   Planning the Project 169

   Organizing Resources and Forming the Team 171

   Leading the Team 174

      KEY TERMS 175
      REVIEW QUESTIONS 175
      PROJECTS 176

PART IV Producing and Distributing Multimedia 179

9 Producing Multimedia Titles 181

Compact Disc 182

   CD Structure 183

   CD Formats 185

The Production Process for CD-ROMs 188

   Premastering 188

   Mastering and Replication 189

   Labeling and Packaging 192

   Production Costs 194

      KEY TERMS 196
      REVIEW QUESTIONS 196
      PROJECTS 197

10 Distributing Multimedia Titles 199

Distributing Multimedia Titles on CD-ROM 200

   Marketing Consumer Titles 200

   Product Strategy 201

   Promotion Strategy 202

   Pricing Strategy 205

   Distribution Strategy 206

   Distribution Alternatives 209

   Marketing Non-Consumer Titles 211

Distributing Multimedia Titles Online 212

Kiosk-based Multimedia 213

      KEY TERMS 214
      REVIEW QUESTIONS 215
      PROJECTS 215

PART V: Multimedia Issues and the Future of Multimedia 217

11 The Internet and the World Wide Web 219

What Are the Internet and the World Wide Web? 220
Multimedia on the World Wide Web 223
    Limitations of the Internet 224
    Developing Multimedia for the World Wide Web 225
    Using the Web as a Source of Multimedia Material 228
    Viewing Multimedia on the Web 231
    Animation on the Web 232
Design Considerations for Multimedia on the Internet 234
    Layout and Features 234
    Hyperlinking 237
        KEY TERMS 239
        REVIEW QUESTIONS 240
        PROJECTS 240

12 Issues and Trends in Multimedia 243

Copyright Issues 244
    Acquiring Rights to Copyrighted Material 245
    Using Material in the Public Domain 245
Privacy Issues 248
Censorship Issues 248
Trends in the Multimedia Industry 249
    The Internet 249
    Hardware 250
    Development 251
        KEY TERMS 252
        REVIEW QUESTIONS 252
        PROJECTS 253

PART VI: Hands-on Tutorial 255

13 Developing an Interactive Presentation: Part 1 257

What Is Action! 3.0? 258
Developing the Exotic Treks Title 262
Working with Palettes and Panels 264
Creating an Action! Scene 268
Creating a Template 270
Adding Text to a Scene 275
Playing a Presentation or a Scene 281
Saving a Presentation 282
Adding a Scene 282
Creating Buttons and Linking Scenes 284
Copying Scenes 288
Creating the Navigation Bar 290
Copying a Group of Objects 296
Adding Text and Linking It to Another Scene 298
Adding Body Text 302
Getting Help While Using Action! 305
Printing a Presentation 307
        REVIEW QUESTIONS 308
        PROJECTS 309

14 Developing an Interactive Presentation: Part 2 311

Hyperlinking Using Embedded Text 312
Adding Graphics with Hyperlinks 314
Adding Motion to Graphics 316
Adding Sound to a Presentation 317
Creating Animations 319
Playing Video and Animation Clips 331
Editing a Movie 334
Working with the Timeline 334
        REVIEW QUESTIONS 339
        PROJECTS 340

Glossary 341   Index 348

PREFACE

**EACHING about multimedia** can be extremely rewarding—it is for me. Helping students learn what multimedia is and how it is changing our lives; helping them understand the technologies—current and forecasted; and working with them in the design and development of multimedia titles has been very satisfying. It also has been an enormous challenge. How do I cover all of the material? How do I keep current in such a dynamic field? How can I *use* the technology to *teach* the technology? These were some of the questions that helped shape the vision for *Multimedia in Action*.

When I first started teaching multimedia several years ago, there were no textbooks developed for the college market. My colleagues and I generated our own materials, drawing on trade publications, software and hardware manuals, industry associations, conferences, and, most important, hands-on experience. Even though there are now dozens of trade books on multimedia, there is still a dearth of textbooks on the subject. Among the reasons for the lack of textbooks are the scope of the subject (multimedia is a vast field, and the challenge is to determine what to include); the fast-changing nature of multimedia, including the impact of the Internet and the World Wide Web; and the need to provide students with a tool they can use to create multimedia titles. Each of these issues is addressed in *Multimedia in Action*.

The textbook provides a comprehensive yet manageable introduction to multimedia, a Web page keeps the content current, and the Action! authoring program provides a hands-on application for the students. *Multimedia in Action* is more than a textbook—it is a teaching package for the instructor and a learning experience for the student.

What Multimedia in Action Covers

This book is designed for those taking an introductory course in multimedia. It helps students understand the practical uses of multimedia as well as how multimedia titles are developed. The book provides a comprehensive study of multimedia, including the following:

- What multimedia is and why it is so important

- Distinguishing between appropriate and inappropriate uses of multimedia

- Identifying the hardware components that make up a multimedia computer and how they are used

- Working with multimedia elements: graphics, animation, sound, text, and video

- Acquiring and producing multimedia elements

- Distinguishing among the many authoring programs and how they are used

- Identifying the steps in creating multimedia titles, including developing the concept, specifying the objectives, creating storyboards, and authoring the title

- Designing both the appearance and the interactivity of a multimedia title

- Discussing the relevant trends and issues surrounding multimedia, such as copyright, privacy, and censorship

- Exploring how multimedia projects are managed

- Identifying the specialists involved in creating multimedia

- Outlining the different CD formats and their uses

- Producing and distributing CD-ROMs

- Exploring the various ways of delivering multimedia, including the Internet

Preface **xi**

Features and Benefits of *Multimedia in Action*

INTERACTIVE CD-BASED EXERCISES

Throughout the chapters students are referred to the *Multimedia in Action* CD-ROM to study practical applications of the concepts they are learning. For example, as the students read about design principles, they are directed to the CD-ROM to study how these principles are used in actual multimedia titles. In addition, the students are presented with a list of questions before viewing the applications. This way the students are studying the applications (and applying what they have read in the book) rather than just viewing them.

To illustrate how the process works, here is an excerpt from chapter 4—Multimedia Elements: Sound, Animation, and Video—followed by screen shots from the CD.

interactive exercise

On the *Multimedia in Action* CD is an example of a virtual reality title. Take a few minutes to view this example.

1. Start the CD.
2. Choose Animation from the contents screen and read the instructions.
3. Click on the questions button and review the questions for Virtual Reality.
4. Click on the demos button and choose Virtual Reality.
5. After viewing the demo, respond to the questions previously reviewed.

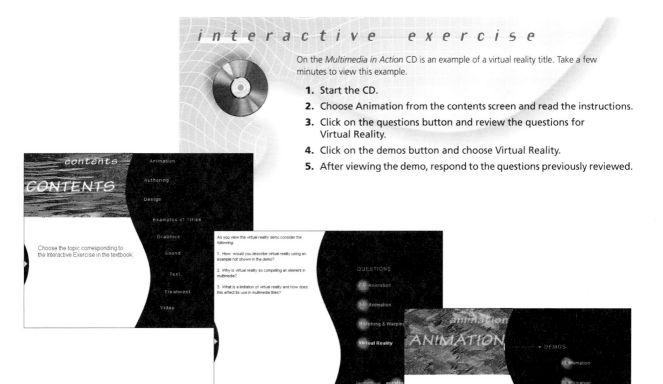

INTERACTIVE WEB SITE

The *Multimedia in Action* Web site provides a way to keep the content current. It includes an instructor section with chapter-by-chapter links to useful Web sites. It also has a student section and a contest for the best title developed using the Action! program. In addition, the site can be used to communicate with the publisher and the author.

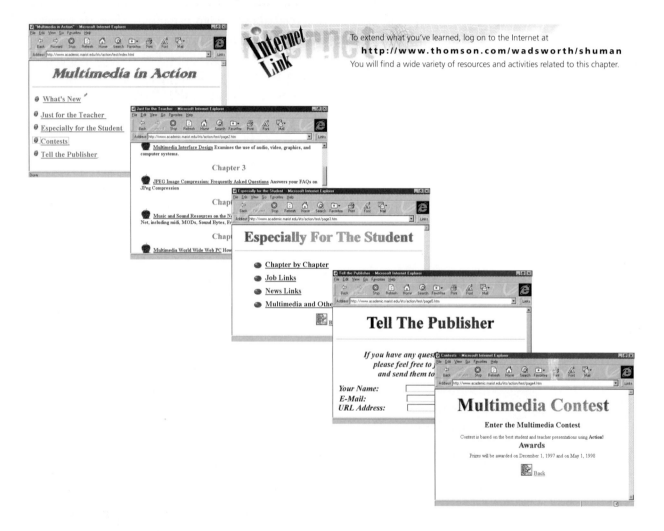

FOCUS ON THE WORLD WIDE WEB

Throughout the book there is a focus on multimedia and the World Wide Web. For example, interactive design issues for multimedia titles delivered via the Web are contrasted with those for developing CD-ROM–based titles; and issues related to the problems of delivering multimedia via the Internet (such as using video) are presented, along with practical solutions such as hybrid (CD-ROM plus the Internet) applications.

Preface xiii

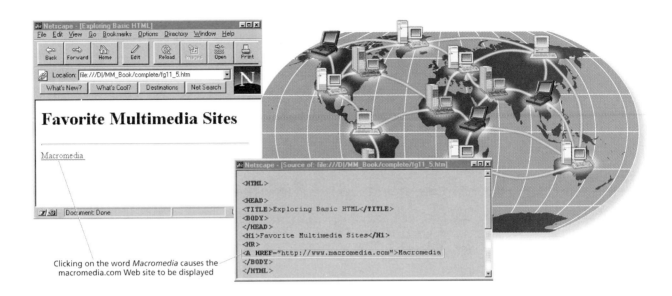

Clicking on the word *Macromedia* causes the macromedia.com Web site to be displayed

Chapter 11 in particular focuses on the Internet and the Web. It details the development, design, and delivery issues related to multimedia and the Web and explains hyperlinking and HTML.

STUDENT-DEVELOPED MULTIMEDIA TITLES

A case study, Exotic Tours, Inc., is the subject matter for a hands-on tutorial that leads students through the development of their own multimedia title. Using a popular presentation development program, Action! 3.0, students learn how to create a multimedia title that includes text, graphics, animation, sound, and video. Extensive use of hyperlinking and special effects add to the learning experience. The Action! program is available to students so that they can practice on their own.

Here are some sample screens from the Exotic Tours multimedia title that the students create.

The Main Menu

An animation

Editing a video clip

STUDENT PROJECTS

Projects are included at the end of each chapter and in the *Instructor's Manual*. These projects provide practical applications to stimulate interest, reinforce learning, and test acquired skills. Several of the projects are designed to help students develop critical-thinking skills through research, analysis, synthesis, and presentation. The following examples are of an end-of-chapter project and an in-class project.

One of the End-of-Chapter Projects for Chapter 11

projects

Use a search engine to visit the following Web sites:

Apple	Microsoft
Asymetrix	MTV
Dorling Kindersley (DK)	Netscape
Edmark	PhotoDisc
Macromedia	Sun Microsystems

Develop a report that includes the following:

- How the company is involved in multimedia (products, services, education)
- Who would be most likely to visit the site
- What value the site can offer the user
- What external links are provided and why
- What techniques are used to address the slow transfer speeds of the Internet
- Which sites are most effective from a visual design standpoint and why
- Which sites are most effective from a navigational design standpoint and why
- Which sites you feel are the most useful and why
- Which sites you feel are the least useful and why

Prepare a brief oral presentation of your report and be ready to present it to your class.

In-Class Exercise for Chapter 1 from the *Instructor's Manual*

Break up the class into groups of three or four students. Give the class a product to sell, a procedure to teach, or a topic to explore with multimedia. Within 30 minutes, the students are to collectively design a multimedia title that meets one of the communication objectives: to sell, to teach, or to inform. They must incorporate each of the multimedia elements into a fictional title. In the remaining time, the students then explain to the class exactly how the communication objectives are met by the title's design.

Other Features and Benefits of *Multimedia in Action*

EXTENSIVE USE OF FOUR-COLOR FIGURES

Concepts and processes are illustrated with more than 190 figures (drawings, photographs, and screen captures) as a way to aid student comprehension, enhance readability, and add interest.

THE TEXTBOOK AS A RESOURCE

The comprehensive nature of the book makes it an excellent reference for students as they take related courses and work in related fields.

TUTORIAL FILES

Graphics, video, and sound files are provided for inclusion in the hands-on tutorial.

GLOSSARY

The glossary provides a quick reference to the key terms used in the book. Key terms within each chapter are shown in ***bold italic*** type in the context in which they are defined. They are listed at the end of each chapter and appear again in the glossary with full definitions.

INSTRUCTOR'S MANUAL

The comprehensive *Instructor's Manual* includes the following features: Chapter Learning Objectives, New Terms and Concepts, Questions to the Class, In-Class Activities, Recommended Multimedia Titles for Review, Selected Bibliography, Transparency Masters, and Test Bank.

Acknowledgments

As is any multimedia project, this book was a team effort. I am indebted to many people for making this textbook a reality. In particular, I would like to thank the following reviewers for their critiques and extremely helpful suggestions: Shihong Steve Chen, Purdue University North Central; John Fodor-Davis, University of Idaho; David John Eatman, Xavier University of Louisiana; Eve Elberg, Pratt Institute; Patricia Freeman, Ohio Northern University; Robert Lucking, Old Dominion University; Gregory X. MacDonald, University of Montana; and Barbara McMullen, Marist College.

Several companies and individuals contributed to the titles included on the *Multimedia in Action* CD-ROM. These include: Amy Gutmann, Edmark Corporation; Chris Brandkamp, Cyan, Inc.; Liz Bethge, Brøderbund Software, Inc.; Chris Jones, Living Books; Ralph V. Giuffre, Humongous Entertainment, Inc.; Joseph Fuller, A.D.A.M. Software, Inc.; Faye Strauss, Sherluck Multimedia Ltd.; and Cheri Grand, DK Multimedia. Others were gracious enough to allow illustrations of their works to be used, including: Joe Burns, Bruce Wolcott, Jennifer Fulton, Reilly Jensen, and Eric Dawes from The Center for Multimedia. I particularly want to express my appreciation to Asha Nelson for her contribution to the creation of the *Multimedia in Action* CD-ROM.

A special thanks also to longtime collaborators Gary Palmatier and Robaire Ream for their extraordinary creative talents in the production of this book, and to Elizabeth von Radics for her meticulous editing skills in making my words make sense. I want to acknowledge my gratitude to Shannon McArdle and the others at Wadsworth Publishing and the Integrated Media Group who contributed to this book. Finally, a heartfelt thanks to my publisher, Kathy Shields, who provided the vision for this project and the management expertise to make it all happen.

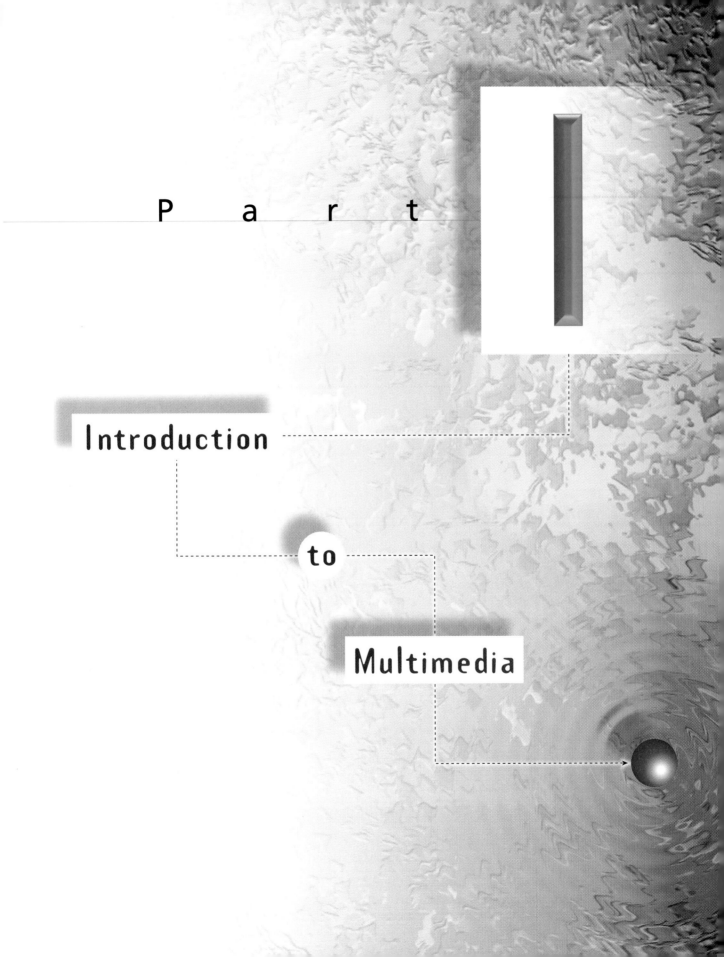

Part I

Introduction to Multimedia

What Is Multimedia?

AFTER COMPLETING THIS CHAPTER YOU WILL BE ABLE TO:

- Define multimedia
- Explain the importance of interactivity
- Trace the growth of multimedia
- Explain why multimedia has become so successful
- List the major categories of multimedia titles
- Distinguish between appropriate and inappropriate uses of multimedia

What Is Multimedia?

PAT, a student at Cascadia College, enters the computer lab, sits down at one of the multimedia computers, and types her password. A menu appears and she selects Biology 101. Another menu appears and she selects DNA Replication. The program asks if she would like to take a pretest, review the process, or begin the tutorial. She selects Begin and watches as a 3-D image of a double helix rotates on the screen. Then the audio narration begins that leads her through the replication process. At any time she can stop the process, review previous steps, ask for help, take a test, or quit the tutorial. Pat takes a moment to reflect on how the textbook readings and instructor lectures make more sense now that she can carry out the process in a virtual lab.

BENNIE pulls his uncle over to the home computer and says, "Want to watch me play my math game?" Uncle Bill watches with fascination as Bennie, a preschooler, solves math problems above his grade level. Numbers flash on the screen, and Bennie quickly performs a calculation and types an answer. After three correct answers, he is rewarded with a game. In the game he must shoot at trash (candy wrappers, empty pop bottles, and the like) that appears in outer space. Each hit causes the trash to be recycled.

CARY stands in front of the kiosk, pressing the buttons on the touch-screen monitor. Each button represents a choice: car model, color, and accessories. After each choice the car as she has configured it is displayed on the screen along with the price. At any time she can take the car for a virtual test drive. When done selecting and equipping her car, she has the kiosk display information on availability, financing, and dealerships.

JOHN is a mechanic for Western Airlines. The company has just purchased five new Boeing 777 airplanes, and he is being trained on how to troubleshoot and repair the new landing-gear systems. He sits at a computer workstation and runs a training module that teaches him how the system works, using graphics, sound, and animation. He navigates inside the landing gear to view each part from every angle. After completing a section, he is tested by being given a repair problem and having to fix it in a simulated environment.

These are actual examples from the education, entertainment, and business fields of what are called multimedia applications or **multimedia titles**. The elements they have in common help us define multimedia. In a generic sense, multimedia is simply the use of many media. Thus, a speaker making a presentation using a slide projector and VCR would be making a multimedia presentation. The word *multimedia* has been popularized, however, as a term that applies to a broad spectrum of *computer-related* products and processes. There are multimedia CD-ROM titles, games, interactive kiosks, CBT (computer-based training) materials, instructional courseware, and online services. What all of these have in common is that they are computer-based; they can incorporate several elements, such as sound and graphics; and they are interactive. Consequently, **multimedia** can be defined as a computer-based interactive communications process that incorporates text, graphics, sound, animation, and video.

A COMPUTER-BASED MEDIUM

Fundamental to the development and delivery of multimedia is a computer capable of incorporating various elements, such as sound and animation, and providing an environment in which the user can interact with the program. Most of us are familiar with desktop and laptop computers such as the Apple Macintosh and PC machines running Microsoft Windows. Models of these computers that are equipped with CD-ROM drives, sound cards, speakers, and sufficient speed and processing power can be used in developing and playing multimedia titles.

INTERACTIVITY: THE KEY COMPONENT

The ability of the user to interact with the program is perhaps the single most critical feature of multimedia. A movie is an examples of a medium that combines numerous elements such as graphics, sound, and animation. But a movie is presented in a linear way—there is a beginning, a middle, and an end. And watching a movie in a theater is a passive process—the viewer has no control. Even when using a VCR (in which the viewer can only control volume; adjust the image quality; and pause, rewind, fast-forward, and stop the movie) the viewer has little control over how the content is presented. Multimedia, on the other hand, allows the content to be presented in a **nonlinear** way, with the user being active rather than passive. Thus, it is **interactive**: The user determines what content is delivered, when it is delivered, and how it is delivered.

Figure 1.1

A flowchart showing how a user controls the information to access in a multimedia title

For example, say you are doing research on Australia, using a multimedia encyclopedia. Figure 1.1 shows a flowchart of the kinds of information you may choose to access from the encyclopedia. You start by typing the word **Australia** as the subject of your search. A main menu appears with several choices, including History, Climate, Economy, Land, and Population, that allow you to determine the content to display. You choose Population,

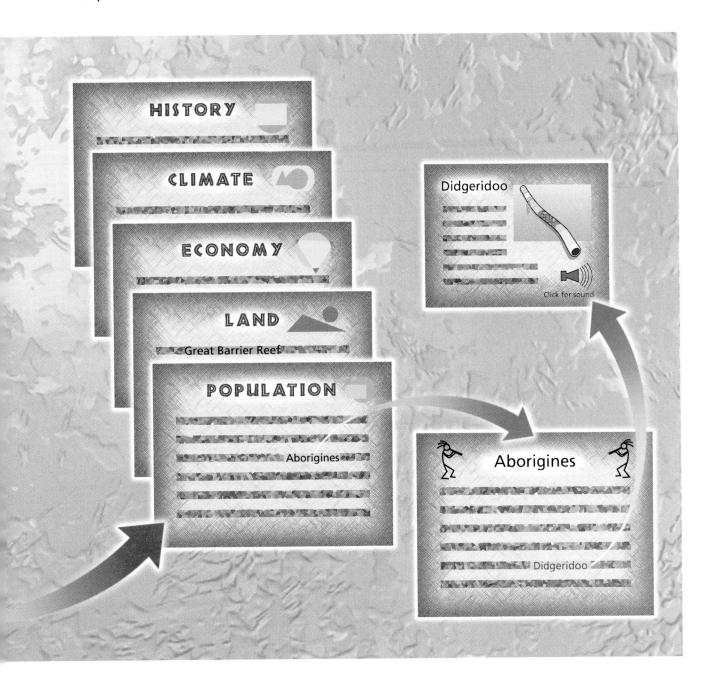

which leads you to an article on aborigines. While reading the article, you notice the word *didgeridoo* in red letters. You click on it, and an illustration and description of the musical instrument appear. You click on an icon resembling a speaker and listen as the didgeridoo is played. You return to the main menu and choose Land and then Great Barrier Reef, and on and on—you have the freedom to explore the topic in any way you choose.

THE ELEMENTS OF MULTIMEDIA

Text-intensive word processing is the most widely used feature of desktop computers. But technological advances now make it possible to use the computer to hear a Martin Luther King Jr. speech, fly a plane in a simulation, display a picture of Saturn, and view a space shuttle launch. All of these elements—text, sound, animation, graphics, and video—can be combined to create a multimedia application.

The Growth of Multimedia

One indication of the phenomenal growth of multimedia is the number of households that own a multimedia computer, that is, a computer with a CD-ROM drive, audio card, and speakers. The chart in figure 1.2 shows that at the end of 1992 less than 1 million households had a multimedia computer. In four years the number of households grew to more than 24 million.

Another indication of the growth of the multimedia industry is the increase in the number of multimedia titles. The chart in figure 1.3 shows how the number of CD-ROM titles increased from approximately 5,000 in 1992 to more than 15,000 in 1996.

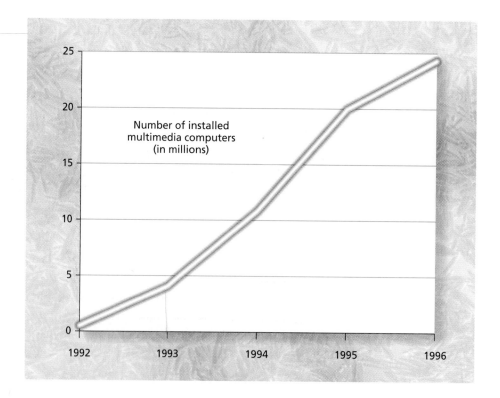

Figure 1.2

Growth of multimedia computers

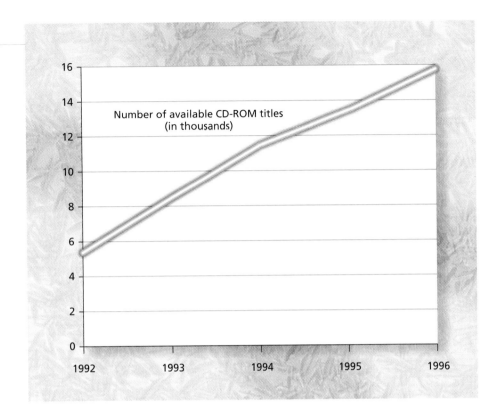

Figure 1.3

Growth in CD-ROM titles

The reasons for the extraordinary growth of multimedia and can be looked at from a marketing and a user standpoint.

GROWTH FROM A MARKETING STANDPOINT

Price Economics tells us that as price declines, demand increases. The chart in figure 1.4 shows how the price of multimedia titles fell from an average of $100 in 1992 to an average of $30 in 1996. During a similar period, the price of new computers capable of playing CDs dropped from around $2,500 to less than $1,500. For people who owned a computer and wanted to upgrade it to play multimedia CD titles (by adding a CD-ROM drive, sound card, and speakers) the price dropped from about $1,000 to less than $300.

Hype The computer industry saw multimedia as the next "killer application." It would do for the home computer market what desktop publishing did for the publishing industry—revolutionize it. The potential size of the industry was compelling. Billions of dollars would be spent on hardware components such as CD players, audio cards, video cards, and speakers, as well as on software programs, such as for authoring, animation, and video and sound editing. Companies geared up to manufacture the

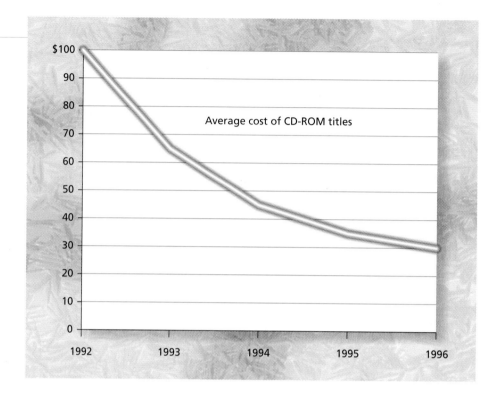

Figure 1.4
Decrease in the price of CD-ROM titles

components and software, trade associations were formed to develop standards and advance the industry, and books and magazines on the subject were published to educate and entertain. Graphic artists, instructional designers, video producers, programmers, and musicians began to learn the new technology that would dramatically affect their professions.

Value added Most personal computers at work and home are used primarily for word processing, a huge improvement over the typewriter, but hardly the best use of the technology. Now, with an investment of a few hundred dollars, the computer can be changed into an entertainment center, educational resource, and marketing tool. Adding value to an existing product is a way to enhance its appeal.

GROWTH FROM A USER STANDPOINT

User control Multimedia empowers the user by giving the user control. And empowerment is motivating. No longer must learning be someone lecturing at us in a mostly one-way, passive process. No longer are we constrained by the limitations of the printed page (instead of reading a printed version in a linear fashion, we can hear the actual person deliver a famous speech and skip around at will). No longer do we have to manipulate (rewind or fast-forward) the VCR to find the desired video clip. The user decides what information to access, when to access it, and the way it will be presented. The user decides if he or she needs to review once more or get additional help.

Individualization   Closely associated with user control is ***individualization***, the ability of a multimedia title to address different learning styles and needs. For example, say you are learning Japanese using a computer-based tutorial. The program might allow you to hear how a word is pronounced by simply clicking on it. You could display the English translation for selected words, or run an animation that illustrates how the words are used in context; you could use a Japanese word processing program to practice your writing skills. You can even use a microphone to record your pronunciation of a Japanese word and compare it with the pronunciation heard on the computer. It is up to the user to decide how the material is presented based on how he or she learns. One person may want to focus on listening skills while another may want to practice vocabulary. Another form of individualization is adjusting the level of difficulty based on user actions. This has long been a feature of video games. In a training environment, the program could be designed to periodically quiz the user to determine how well the material is being learned. Based on the user's answers, the program would branch to easier or more difficult lessons—individualization.

Action   Reading a book, listening to a lecture, and watching a videotape are somewhat passive processes. Using a computer to play a flight simulator game or dissecting a frog in a virtual lab are active processes. Consider this: You are using a multimedia title to study a Shakespearean play. You begin by reading a passage from the play, then you choose to see the passage acted out by viewing a digitized video clip of an actual performance. Finally, you decide to test your knowledge of the passage by taking a quiz—and getting immediate feedback. Based on the feedback, you decide to review the material again or go on to a new passage—an active process.

Examples of Multimedia

Note: The *Multimedia in Action* CD that accompanies this book has several examples taken from actual multimedia titles. You will be directed to view the examples at various points in this chapter and throughout the rest of the book. Look for the ***interactive exercise*** heading; its first occurrance is on the following page.

The Major Categories of Multimedia Titles

It is useful to distinguish between different categories of multimedia, because creating a multimedia title may vary depending on its category. For example, the use of animated cartoon characters might be appropriate for a multimedia game, but inappropriate for a marketing presentation. There are various ways to group multimedia titles. They could be classifed by market, such as home, business, government, and school; or by user, such as child, adult, teacher, and student; or by category, such as education, entertainment, and reference. Moreover, many titles could fall into more than one group. For example, a title that teaches a child how to spell by using cartoon characters and providing rewards may be considered a game by the child and an educational program by the parent.

For some of the following categories, a sample title can be found on the *Multimedia in Action* CD. These titles are recognized as leaders in their respective categories. As you read through this section, take a few moments to study each title to determine why it has been successful.

ENTERTAINMENT

Game developers were pioneers in the use of multimedia. From large arcade-style video games to handheld Nintendo Game Boys, the focus has been on action and graphics. The developer needs to attract, engage, captivate, and challenge the user. Multimedia developers have taken the emphasis from pure action to action plus storytelling, from games to entertainment, from the physical (hand/eye coordination) to the mental (solving the mystery,

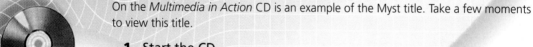

interactive exercise

On the *Multimedia in Action* CD is an example of the Myst title. Take a few moments to view this title.

1. Start the CD.
2. Choose Examples of titles from the contents screen and read the instructions.
3. Click on the questions button and review the questions.
4. Click on the demos button and choose Myst.
5. After viewing the title, respond to the questions previously reviewed.

overcoming evil, outwitting the opponent). Multimedia can incorporate the fast action, vivid colors, 3-D animations, and elaborate sound effects that are essential to entertainment. It can also provide the rewards, recognition, and sense of accomplishment that are often a part of entertainment titles. 7th Guest is an interactive mystery in which you search for the secret of the 7th Guest in a realistic 3-D haunted house. It includes full-motion video of live action. Another successful multimedia title is Myst. This game is an adventure that begins on an island and uses exploration as a way for the player to experience the mysteries of the island.

EDUCATION

A goal of the educator is to facilitate learning—to help the student gain a body of knowledge, acquire specific skills, and function successfully in society. One of the greatest challenges to an educator is the diversity of students, especially in the different ways they learn. Multimedia can accommodate different learning styles: Some students learn better through association, others by experimentation; some are more visually oriented and others are more auditory. Multimedia can present material in the way we think—in a manner that is nonlinear. It lets us review specific aspects as often as we like, skipping around as necessary. It is motivating, as it allows the user to take charge of his or her learning. Multimedia can provide feedback, adjust the level of difficulty, and evaluate skills. And it can make learning fun. A.D.A.M. is a multimedia anatomy program that includes extensive structure labeling, indicating, for example, specific branches of veins and arteries, the entire lymph system, and individual ligaments. It includes more than 18,000 structure identification labels.

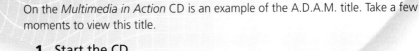

interactive exercise

On the *Multimedia in Action* CD is an example of the A.D.A.M. title. Take a few moments to view this title.

1. Start the CD.
2. Choose Examples of titles from the contents screen and read the instructions.
3. Click on the questions button and review the questions.
4. Click on the demos button and choose A.D.A.M.
5. After viewing the title, respond to the questions previously reviewed.

CORPORATE COMMUNICATIONS

Marketing and training The ultimate goal of a marketer is to sell a product, service, or idea, usually through informing and persuading. First, however, the marketer must attract attention to a message. Multimedia can accomplish this through the use of sound, animation, and graphics and through addressing specific needs of a target audience. Companies now distribute their product catalogs on CD-ROM, allowing the buyer to customize the product (change the color, add accessories, and so on) and then order it. Certain magazines are now published and distributed on CD rather than in print. Touch-screen kiosks are strategically placed in shopping malls, retail stores, and car dealerships. Home pages on the World Wide Web provide interactive, online shopping via the Internet. One example of marketing and multimedia is The Merchant, which contains more than 20 catalogs including J.C. Penney, L.L. Bean, Spiegel, and Target. It allows you to browse through a catalog and select the merchandise to view, including size, color, and other information. It also displays the price and ordering procedure. The Merchant even includes sound clips of music CDs.

Presentations and training Thousands of *multimedia presentations* are made in the business world every day. Company CEOs give the annual report to a meeting of stockholders. Sales reps pitch their product line to a group of potential customers. A conference keynote speaker tells an audience about industry trends. From an electronic slide show to an interactive video display, multimedia can enhance presentations. Multimedia gives the presenter a tool to attract and focus the audience's attention, reinforce key concepts, and enliven the presentation. Another form of corporate communications involves training employees using multimedia-enhanced materials. TRW, a large Ohio-based corporation, developed an interactive employee orientation CD that includes company goals, history, products, and codes of conduct. The interactive CD, which includes video segments, provides an effective way for TRW to introduce new employees to the company.

REFERENCE

Encyclopedias, census data, Yellow Pages directories, and dictionaries are examples of CD reference titles. In many cases these are electronic versions of reference books. The challenge to the developer is to make it easy for the user to find the desired information, and to effectively use other multimedia elements such as sound, video, and animation. Microsoft Encarta includes the 29 volumes of the *Funk and Wagnalls* encyclopedia. Encarta has more than 100 animations and video, 9 hours of sound, and 350 music segments. It also includes a timeline that helps you see how epoch civilizations and events relate to each other. When a subject interests you, click on it to access more information. It also has a browse function to allow you to skim through thousands of topics.

interactive exercise

On the *Multimedia in Action* CD is an example of a reference title. Take a few moments to view this title.

1. Start the CD.
2. Choose Examples of titles from the contents screen and read the instructions.
3. Click on the questions button and review the questions.
4. Click on the demos button and choose Reference.
5. After viewing the title, respond to the questions previously reviewed.

Other Categories of Multimedia

Edutainment As the name suggests, *edutainment* is the combination of education and entertainment. Many multimedia titles, especially children's games such as Math Blaster, fit this category.

Training Multimedia titles that focus on developing specific skills usually related to a particular job are called training titles. The Boeing company has an entire training division dedicated to developing multimedia titles that instruct mechanics and pilots on new aircraft systems. Holiday Inn uses a computer-based interactive role-playing game to improve customer service.

Recreation Hobbies and sports are examples of the types of titles that could be classified as recreation. These often give the user a vicarious experience

Figure 1.5

Examples of how CD titles can fit into more than one category

Multimedia Title	Entertainment	Education	Reference	Recreation
7th Guest (animated adventure)	●			
Encarta (encyclopedia)		●	●	
Bookshelf (reference titles)		●	●	
Just Grandma and Me (children's story)	●	●		
MS Golf (game)	●			●
A.D.A.M. (anatomy instruction)	●	●	●	
From Alice to Ocean (trek across Australia)	●	●		●

such as being able to "play" the most famous golf courses in the world or simulate flying over 3-D cityscapes.

Figure 1.5 lists the categories of multimedia and shows how selected titles can fit within more than one category.

Delivering Multimedia

Until now we have focused on the way in which multimedia titles can be classified. In this section you will learn about different ways to deliver multimedia. Chapter 10, "Distributing Multimedia Titles," covers this topic in greater detail.

Chapter 1 *What Is Multimedia?*

Figure 1.6
Compact discs

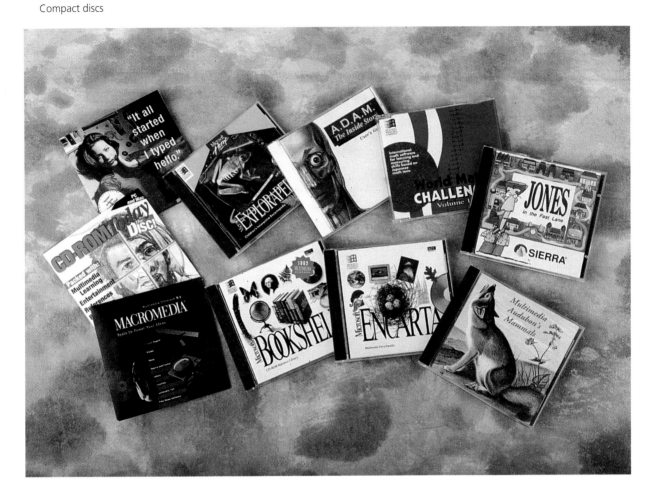

COMPACT DISC

The growth of multimedia, as shown earlier, is often expressed in terms of the growth of ***compact disc (CD)*** titles (see figure 1.6). CDs are a popular medium because they can hold substantial amounts of data, including sound and video. They are relatively inexpensive; easy to mass produce, distribute, and transport; and they take up little retail shelf space. In addition, it is now common for computers to be sold with CD-ROM drives, sound cards, and speakers.

KIOSK

A ***kiosk*** is a stand-alone or networked computer system that allows the user to access information, perform transactions, and even play games (see

Figure 1.7

A kiosk

figure 1.7). Examples are college informational kiosks that students use to learn about academic programs, print out schedules and transcripts, and access a campus map; retail store kiosks that allow customers to locate merchandise, print out coupons, and purchase products; and museum kiosks that allow the user to locate specific works of art, view parts of a collection that are not on display, and obtain detailed information about the artists. Kiosks are useful in disseminating information especially in high-traffic areas, providing value-added services to customers (convenience), and reducing personnel costs. They are expensive because of the investment in hardware and require continual updating of the content.

ONLINE

One of the fastest-growing areas for multimedia delivery is ***online***, which includes telecommunications and the Internet (see figure 1.8). ***Telecommunications*** involving phone line, satellite, and cable transmission is being

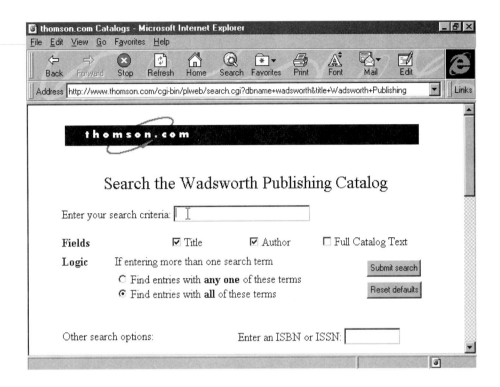

Figure 1.8
An online catalog

used by educational institutions to deliver multimedia courseware to rural areas, and by companies for teleconferencing and training. The use of the ***Internet*** is growing exponentially. Companies are developing home pages for the ***World Wide Web (WWW)*** that allow customers to purchase products, access product information including video demonstrations, and even subscribe to real-time multimedia events such as rock concerts.

Inappropriate Use of Multimedia

Despite the advantages, it is not always appropriate to use multimedia. The following section outlines some considerations.

TEXT-INTENSIVE CONTENT

Reading large amounts of text on a computer screen is tedious and tiring, both physically and mentally. Placing a book on a CD and expecting the user to read it "from cover to cover" is not realistic. Developing interactive books, on the other hand, in which the user becomes an active participant and can make choices that affect the story line and outcome can be effective. Similarly, multimedia reference titles can contain a great deal of text; but, by allowing the user control over the content and by adding other elements (such as animation, sound, and video), they can overcome the drawbacks of being text intensive. Chapter 3 has an in-depth discussion of the use of text in multimedia titles.

LINEAR CONTENT

Soon entire full-length movies will be distributed on CD. But watching a movie or any digitized video from beginning to end is not multimedia. Here is an example of an inappropriate use of video: A company wanted to showcase its high-tech image by sending out invitations on CD to an upcoming conference. The content of the CD was a well-developed, 10-minute videotape about the company, which included interviews, product demos, future plans, financial data, and so forth. The result was a 10-minute video played in a small window on a computer screen with poor resolution and no user control. How many potential conference attendees will have multimedia playback computers? Will the user want to watch 10 minutes of poor-quality video on a computer screen? Will the user be impressed with the company's attempt at creating a "high-tech" image? A better approach would be to allow the user to choose from a menu that includes company background, interviews, product demonstrations, and so on, and then play a short video clip appropriate for the choice.

Some Cost-effective Alternatives

Although the allure of multimedia is substantial, developers must weigh the development time and the costs of alternatives. Can the communication objectives be accomplished more effectively and/or inexpensively using another process? Is there time to develop a compelling title? Could a presenter simply use overhead transparencies? Could a company distribute videotapes (virtually everyone has access to a VCR) rather than CDs to deliver its message?

Even with all of its advantages, multimedia cannot make up for a lack of content, poor design, targeting the wrong audience, or delivery by a mediocre presenter.

To extend what you've learned, log on to the Internet at

http://www.thomson.com/wadsworth/shuman

You will find a wide variety of resources and activities related to this chapter.

Chapter 1 What Is Multimedia?   **21**

key terms

compact disc (CD)
edutainment
individualization
interactive
Internet
kiosk
multimedia

multimedia presentation
multimedia title
nonlinear
online
telecommunications
World Wide Web (WWW)

review questions

1. Define multimedia: _COMPUTER BASED, INTERACTIVE_
   COMM PROCESS   TGSAV
   _____
   _____

2. The elements that can be combined to create a multimedia title are:
   a. _T -_
   b. _G -_
   c. _S -_
   d. _A -_
   e. _V -_

3. (**T**) **F** More than 20 million households have computers that can play CD-ROM titles.

4. Three reasons for the growth of multimedia from a marketing standpoint are:
   a. _P_
   b. _H - Revolution, Potential_
   c. _V - EC, Edu, Shipping, mkt_

5. Three reasons for the growth of multimedia from a user standpoint are:
   a. _Uc_
   b. _I - Styles, need_
   c. _A - Interactive_

6. List three categories of multimedia titles:

a. Edu

b. Ref

c. Ent

7. (T) F It is not unusual for one CD title to fall into more than one category.

8. T (F) An advantage of an informational kiosk is that the content does not require updating.

9. List two considerations for the inappropriate use of multimedia:

a. Reading text on computer

b. Reference on text-intensive

10. List two considerations for the appropriate use of multimedia:

a. Interactive / Individualized / Action pre

b. User controlled ref

projects

1. Submit a two-page (typed, double-spaced) review of a magazine or newspaper article having to do with multimedia. Include the following:

- Name and date of publication
- Name of article
- Review of article (in your own words)

Interview a person who is currently working with multimedia and submit a (typed, double-spaced) summary of the interview. Include the following:

- Name of person interviewed
- Company, title, duties
- How he/she is using multimedia

Develop a five-minute presentation based on the preceding review or interview and be prepared to deliver the presentation to the class.

2. Choose a CD multimedia title (if possible, in a lab or one you have at home or work) and develop a report that includes the following:
   - Purpose of the product
   - Intended audience
   - Categories the title would fit in and why
   - Elements of multimedia used (sound, animation)
   - Design of the initial screen
   - Information contained on the initial screen
   - Method of navigation through the program
   - Approximate time needed to complete the entire product
   - How the product is divided if there is not time for the user to complete all of it
   - Strengths of the product
   - Weaknesses of the product
   - How you would change the product to make it better
   - Cost of the product

3. Find a kiosk and use it. In a typed, double-spaced, two-page report, describe the following:
   - Location of the kiosk
   - Purpose of the kiosk
   - Type of information (if appropriate)
   - Intended audience
   - How the user works with the kiosk (touch screen, keyboard, sound, pen, mouse)
   - Navigation process, that is, how the user makes selections (menus, buttons, graphics)
   - Multimedia elements used (animation, sound, video, text, graphics)—give examples
   - Equipment (if possible, make of computer, color, printer, CD, videodisc, portable)
   - Other (who makes it, who developed it, who maintains it, cost)

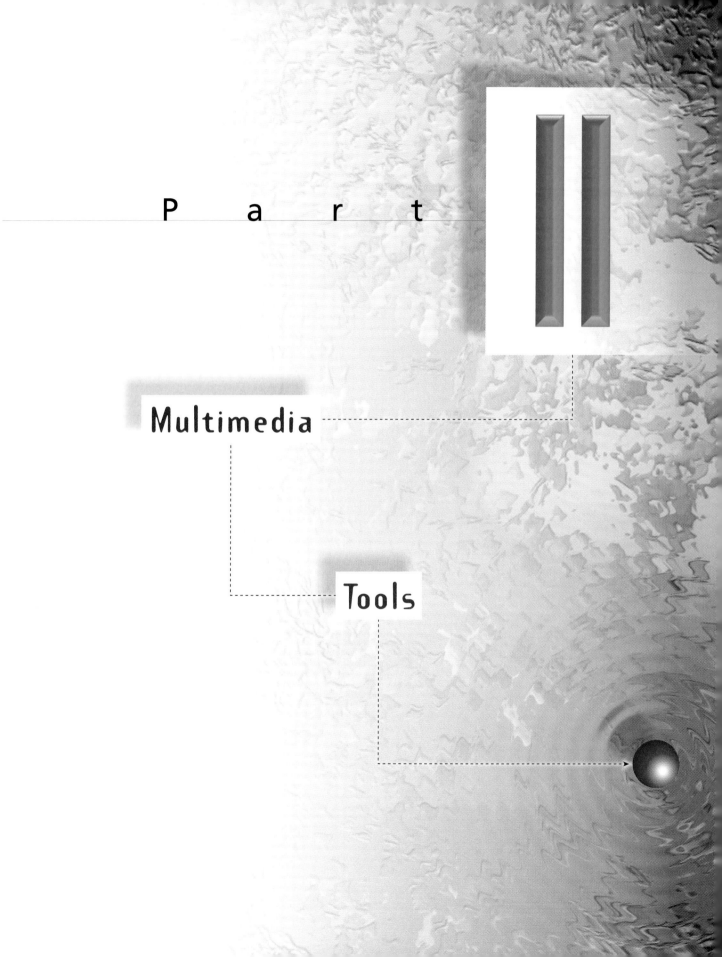

Hardware Components of a Multimedia System

AFTER COMPLETING THIS CHAPTER YOU WILL BE ABLE TO:

- Describe why hardware standards are important

- Describe the MPC standard and specify its significance

- Describe the multimedia hardware components for playback and development systems

- List and describe the major multimedia hardware peripherals

WHEN studying multimedia computer systems, it is useful to distinguish between those systems used for development and those used for playback. **Development systems** need to be the fastest and most powerful, and need to have the largest storage capacity that a company can afford. The development process is people intensive, and people's time costs money. The better tools they have, the quicker they can work. On the other hand, the type of **playback system** is related to the installed market, that is, the computers in use. When multimedia was developing in the 1980s, there were millions of desktop computers in businesses, homes, and schools. As the technology was improving to allow animation, sound, and video to be played on a desktop computer, dozens of companies became interested in developing multimedia titles. They were faced with a dilemma: There were so many different types of systems, varying in speed, capacity, and display capabilities, that developers risked creating a title that would run on only a limited number of computers. For example, should they use 16 colors or 256 colors for their graphics? There might be more computers that run 16 colors, but the image quality would be significantly less. Developers were therefore hesitant to invest the hundreds of thousands of dollars it might take to create a title. This lack of standardization was to many developers a barrier to entering the industry. Another barrier was the lack of a way to deliver, through the desktop computer, the large amount of data required for a multimedia title. The first barrier was addressed by the development of industry standards. The second barrier was overcome by the widespread use of compact discs.

You learned in chapter 1 that multimedia can incorporate text, graphics, animation, sound, and video. With the exception of text, these elements can be very large in size. For example, 15 seconds of compressed video could take more than 15 **megabytes** (15 million characters of data, or 15 MB) of storage on a disk. Thus floppy disks, which hold less than 2 MB of data, are not capable of providing the necessary storage capacity for multimedia titles. The computer industry understood that in order to deliver multimedia titles to the desktop, some medium needed to be developed that would hold hundreds of megabytes of data. Because a CD can hold approximately 650 MB of information, it became the medium of choice. Figure 2.1 shows a typical configuration for a multimedia-capable computer used for playback. From the outside it looks similar to any desktop computer with one exception: a CD-ROM drive.

Figure 2.1

A typical configuration for a multimedia-capable computer

The Multimedia Personal Computer

Throughout the history of personal computers, it has become evident that hardware standards are important. One of the reasons for the recognized quality of Macintosh computers is that Apple Computer, Inc., controlled the specifications of each of its models. Thus, developers creating programs for the Macintosh, including multimedia titles, knew exactly what hardware components were involved. Unfortunately, the installed base of Macintosh computers is relatively small (10 percent of the market) compared with Windows-based machines (80 percent of the market). Windows-based computers were being manufactured by many companies and with many different standards. Windows-based computer manufacturers realized that for the industry to grow, customers needed computers that could do more things, including running multimedia titles. Without standardization, however, developers had little incentive to invest in the creation of titles.

Figure 2.2
MPC logos

Thus in 1990 a group of companies agreed on the ***Multimedia Personal Computer (MPC) specifications*** for Windows-based machines. These became know as the Level 1 specifications. An MPC logo was established, and those developers who sold hardware that met the specifications or software that ran on the MPC machines were allowed to use the logo (see figure 2.2).

In 1991 the Multimedia PC Marketing Council was formed to promote and revise the standards. This council was affiliated with the Software Publishers Association, the predominant industry trade group, and included Microsoft, IBM, Philips Consumer Electronics, Compton's NewMedia, and NEC Technologies. Those developing the standards were faced with a trade-off: size of the market versus power of the computer system. If the standards were set high, developers could create more-exciting and more-compelling titles, increasing the market appeal. But the installed base of computers were those at the low end, and consumers were hesitant to purchase entirely new systems. In 1993 the Level 2 specifications were published.

The results of these standards on the industry were dramatic. If you were a developer, the risks involved in creating titles were substantially reduced. You knew that if you created a CD title that met the MPC specifications, the title would run on all MPC machines. In addition, when you put the MPC logo on your product, customers would know that your title would run on their MPC machines. In 1995 the Software Publishers Association took responsibility for upgrading the MPC standards and released the MPC 3 specifications. MPC 3 included a requirement for video compression that allows full-screen and full-motion video. Figure 2.3 compares the minimum requirements for Levels 1, 2, and 3.

The MPC specifications focused on the speed and capacity of the system unit, the resolution and colors for the display unit (monitor), and the quality of the CD-ROM drive and sound card.

Figure 2.3

Levels 1, 2, and 3 MPC minimum specifications

	Level 1	**Level 2**	**Level 3**
RAM	2 MB	4 MB (8 MB recommended)	8 MB
Processor	386SX 16 MHz	486SX 25 MHz	Pentium 75 MHz
Hard drive	30 MB	160 MB	540 MB
CD-ROM drive	Single-speed, 150 Kbps transfer rate; max average seek time: 1 second	Double-speed, 300 Kbps transfer rate; max average seek time: 400 milliseconds	Quad-speed, 600 Kbps transfer rate; max average seek time: 200 milliseconds
Sound	8-bit digital sound	16-bit digital sound	16-bit CD quality
Video display	640 × 480 pixels; 16 colors (256 recommended)	640 × 480 pixels; 65,536 colors	640 × 480 pixels; 65,536 colors

The Playback System

PROCESSOR

The type of ***processor*** determines, among other things, how quickly data is processed and transferred. This becomes critical as the multimedia title becomes more graphic-intensive. The MPC Level 2 standard is a 486SX 25 MHz processor. A minimum Macintosh system would be a model LC 68030 with a 25 MHz processor. When a multimedia title is purchased, its package will specify a minimum configuration. The lower the configuration, the larger the potential market. Often the package will also specify a recommended configuration that will provide optimal-quality video, sound, and graphics and smoother animations.

MEMORY

The two basic types of memory in a computer are temporary and permanent. The temporary memory, called random-access memory, or ***RAM***, is used to store instructions and data that are used while an application is running. For example, if you are playing a game on the computer, some of the game instructions would be loaded into the temporary memory—RAM—as the game is played. When you turn off the computer when you are finished, the instructions are erased from the temporary memory.

The computer's hard drive is used to permanently store program instructions that are needed each time the program is run. This is called read-only memory, or **ROM**. The process works like this: Say you purchase a CD title called Interactive Skiing. At home you insert the CD into the drive and follow the instructions to install the program. A set of instructions is copied from the CD to the computer's hard drive. These instructions allow you to run the program without having to repeat the installation process. Thus, the next time you want to play the game, you simply insert the CD and select the program's icon from the screen. While the program is running, instructions and data are loaded from the hard drive and CD as needed. Although the term *permanent* is used to describe the read-only memory, you can, of course, erase programs from a hard drive.

MONITOR AND VIDEO CARD

The ***monitor***, or display, is critical to the playback system, because it provides the primary feedback to the user. Standards have been established for the screen resolution and the number of colors.

Figure 2.4

The letter *P* made up of pixels on a grid

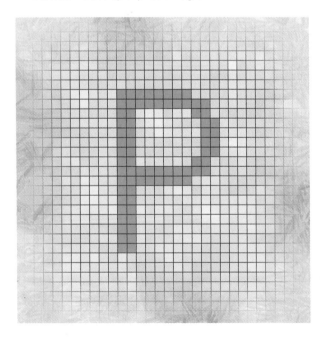

Screen ***resolution*** is measured in the number of dots displayed on the monitor. These dots are called ***pixels*** (short for picture elements) and are the smallest units a monitor can display. Figure 2.4 shows how the letter *P* is made up of multiple pixels on a grid. The more pixels, the sharper the screen image. A standard resolution is 640 pixels across and 480 down the screen. Figure 2.5 shows the same image using two different screen resolutions, 640 × 480 and 1024 × 768. Notice the better quality of the image displayed in 1024 × 768 resolution. The number of pixels is determined by the video graphics adapter card and its memory capacity. For Windows-based computers, the VGA (Video Graphics Array) cards support a resolution of 640 × 480, whereas the SVGA (Super VGA) cards can support much higher screen resolutions.

The ***video card*** also determines the number of colors that can be displayed on the screen and,

Figure 2.5

The same image with two different screen resolutions, 640 × 480 and 1024 × 768

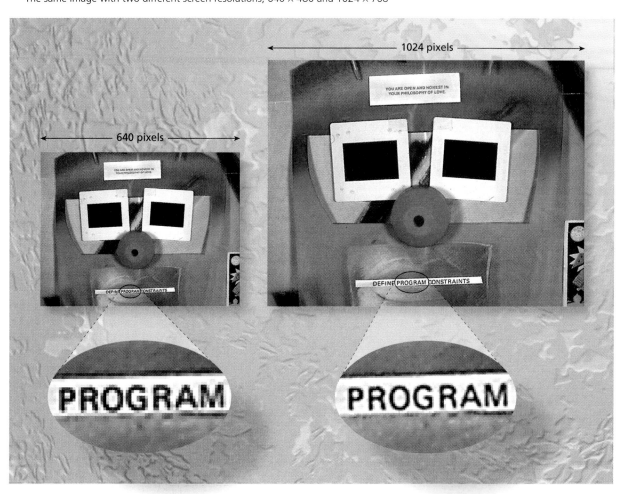

hence, the color quality: The more colors, the higher the image quality. Each pixel can display from one color to millions of colors. The number of colors is determined by information associated with the pixel. The information is coded in *bits* (binary digits). This is how it works: Computers operate on the basis of the flow of electricity and the sensing of electrical impulses. The electrical impulse is either present or not present. Symbols used to represent this are *1* for present and *0* for not present. Because there are two symbols, this is called a ***binary*** system. Everything that is done by the computer can be represented by this binary system. This includes determining the number of colors that can be displayed by a pixel. Let's say that you want to use 16 different colors in your graphics. Each pixel would need to be able to display each of the 16

Figure 2.6

How 4 bits allow for 16 colors

Bit Values				Pixel Color
8	4	2	1	
0	0	0	0	0
0	0	0	1	1
0	0	1	0	2
0	0	1	1	3
0	1	0	0	4
0	1	0	1	5
0	1	1	0	6
0	1	1	1	7
1	0	0	0	8
1	0	0	1	9
1	0	1	0	10
1	0	1	1	11
1	1	0	0	12
1	1	0	1	13
1	1	1	0	14
1	1	1	1	15

different colors. Each color could be assigned a number from 0 to 15 (16 total colors). So the question becomes: How many bits are needed to represent 16 colors? Using a binary number system, 4 bits in different combinations of "on" and "off" can represent 16 different numbers (colors). The process requires assigning each bit a value based on its position relative to the other bits. This is shown in figure 2.6. The *8, 4, 2,* and *1* represent the values associated with each bit. The 0's and 1's indicate whether the bit is turned on. The sum of the bit values that are turned on

Figure 2.7
Different screen resolutions, bits, and colors available with different video cards

Screen Size in Pixels	Color Depth in Bits	Number of Colors	Video Graphics Adapter
640 × 480	4	16	VGA or SVGA
640 × 480	8	256	SVGA
800 × 600	8	256	SVGA
1024 × 768	16	65,536	SVGA
1280 × 960	24	16.7 million	SVGA

represents the number associated with the pixel color. To use more than 16 colors, you need to increase the number of bits as shown in figure 2.7. The preceding example is used to illustrate the relationship between the binary system and the display of different colors. In practice, 16 colors would be too few to provide high-quality images. In most cases 256 colors would be a minimum.

AUDIO CARD

Whether in games that provide sound effects or education titles that teach foreign languages, sound can be an extremely important element in a multimedia title. In order to incorporate sound, the computer needs an audio card, such as Sound Blaster, and speakers. Just as a video card is used to display digital images, an *audio card* is used to play digital sounds. Also, just as the number of bits used to represent an image determines the quality of the image, the number of bits used to represent a sound determines the quality of the sound. The standard is now 16-bit sound. In order to be played through a computer, sounds need to be digitized. The process, called sampling, changes an analog signal into a digital signal. Sampling is explained in chapter 4.

CD-ROM DRIVE

CD-ROM stands for compact disc–read-only memory. A **CD-ROM drive** reads the data (graphics, sound, and text) on the CD and transfers it to the computer. If the transfer time is too long, the user may get impatient. The CD-ROM drive determines the type of CD that can be played, the speed at which data is located on the CD, and the speed at which data is transferred from the disc to the computer. The data transfer rate is measured in kilobytes per second, or Kbps. A *kilobyte (KB)* is roughly equivalent to 1,000 characters of data. The first CD drives were classified as single-speed with a transfer rate of 150 Kbps. Now there are quad (4×), as well as 6×, 8×, and even 12× speed drives with a transfer rate of 1800 Kbps. The Level 3 MPC standard is a quad-speed drive with a transfer rate of 600 Kbps. This compares to hard disk drives that transfer 13 MB (megabytes, or million bytes) per second. The seek time required to find a specific piece of data on the CD is measured in *milliseconds (ms)*, that is, thousandths of a second. The Level 3 standard is 200ms. This compares to hard disk drives that have seek times as low as 10ms.

UPGRADE KITS AND BUNDLED SYSTEMS

In order to take advantage of the installed base of computers, several manufacturers produced *upgrade kits* that could take a basic desktop computer and change it into an MPC system. These kits contained a CD-ROM drive, audio card, speakers, and the software needed to run those components. Initially, the price of these kits was relatively high. But now, because of increased demand and lower production costs, prices start at less than $100. Figure 2.8 shows a multimedia upgrade kit.

Figure 2.8

A multimedia upgrade kit

The Development System

It is possible to develop multimedia titles using MPC Level 2 (or Macintosh equivalent) computer systems. Developers realize, however, that in order to produce commercial titles that can compete in terms of high-end graphics, sound, and video, they need to invest in the best equipment they can afford. Not only will high-end equipment provide the necessary quality, it can lower production costs by reducing the time programmers, graphic artists, animators, and others spend in creating their part of the title. In addition to the computer system, several related hardware components are needed in the development process. Figure 2.9 shows a typical development system.

Both Windows-based and Macintosh platforms are used in creating multimedia titles. But because the Macintosh was the first popular computer to provide a graphical interface and because it had superior handling of graphics and cross-platform capabilities, it has been used extensively in multimedia development. Fortunately, software is available that allows a developer to choose either platform for creating a multimedia title and have the title play back on both platforms.

Figure 2.9

A typical development system

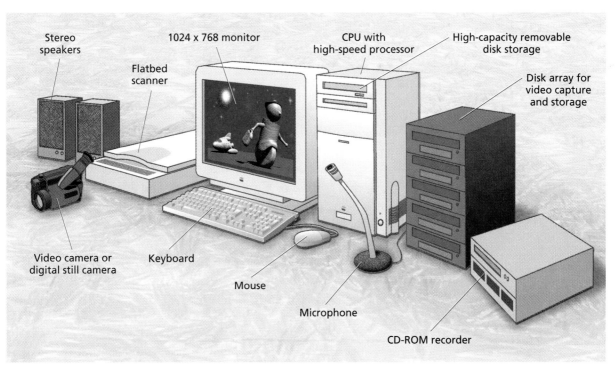

PROCESSOR

Today a model 486DX 66 MHz unit would be considered a minimum for development work. A Pentium 100 or better would be desirable. A Macintosh with a 68040 processor running at 33 MHz would be a minimum configuration, with a Power Mac with more than 100 MHz being desirable.

MEMORY

Multimedia titles are extremely memory intensive. Therefore, 16 MB of RAM would be a minimum, with 32 or 64 recommended. Hard drive disk space is determined by the number of programs that need to be stored on the drive. It is common to use several programs to create a multimedia title, including drawing, authoring, and image, sound, and video editing. All of these take up disk space. Add to this the space required for the various elements of the title and you soon run out of room. Minimum hard disk space would be 1 *gigabyte* (1 billion bytes, or 1 GB).

VIDEO CAPTURE CARD

If video is going to be used, it needs to be digitized using a video card. The card fits internally within the computer, and a video source (camera, VCR, TV, videodisc) is plugged into the card. As the video signal is sent from the source, it is captured, digitized, and stored. Later it could be edited by deleting frames, adding text, adding sound, and so forth. The process for digitizing video is presented in chapter 4.

MONITOR

Whereas playback units typically have 14-inch or smaller monitors, developers need larger (20-inch) models. This allows them to enlarge an image for detailed editing and to use the monitor as you would a desktop, with several items displayed and available. Ideally, the developer would have two monitors, one that is used to work on the title, and another for displaying the title as it is being created.

PERIPHERALS

In addition to the basic computer system, other hardware —called *peripheral* devices, or peripheral—may be needed when developing a multimedia title.

Scanner A *scanner* is used to create digitized images so they can be incorporated into multimedia titles. Scanners are an excellent way to generate

Figure 2.10

A comparison of selected scanners

Make	Mac	PC	Colors	Cost
Hewlett-Packard	✓	✓	>1 billion	$ 900
Cannon		✓	millions	$ 750
Scan Maker III	✓		68.7 billion	$ 2,000

graphics from photographs, books, and artwork—essentially any printed material as well as any object that can be placed in a scanner. Scanners vary in terms of configuration (flatbed, handheld, slide), quality (number of colors), editing features (such as the ability to zoom in on an image and crop an image), editing features (such as adjusting color, contrast, and brightness), and price. The typical development system shown in figure 2.9 includes a scanner; figure 2.10 shows a comparison of various scanners.

External storage Additional storage space to relieve the pressure on a computer's hard drive is provided by *external storage* devices, like the one shown in figure 2.11. They can be used to back up data from the hard drive, which prevents accidental loss of work. In addition, they give needed portability when files must be transferred from one system to another and

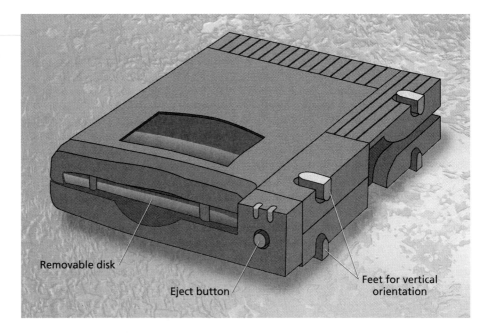

Figure 2.11

An external storage device

Figure 2.12

Selected external storage devices

Make	Mac	PC	Capacity	Cost
Syquest	✓	✓	135 MB	$ 240
Iomega Zip	✓	✓	100 MB	$ 150
Bernoulli	✓	✓	230 MB	$ 470
Quantum External Hard Drive	✓	✓	4.3 GB	$ 1,100

the computers are not networked together. For example, if you develop a multimedia title and need to transfer it onto a CD, the title could be placed on an external storage device which would be physically delivered to a company that masters CDs. Figure 2.12 compares some of the different external storage devices.

CD recorder As the price of CD recorders has dropped from $10,000 to less than $500, multimedia developers are using them in multiple ways. CD recorders can be used to create single CDs of products for testing, to generate finished CDs when only a few are required, to provide a way to deliver the files to a manufacturer for mass production, and to back up and archive data.

Digital camera A *digital camera* is used to capture still images just like a regular camera (see figure 2.13). After taking a picture, you can transfer it directly from the camera to the computer as a graphic image. Software that comes with the camera allows you to edit the image and save it as a graphic file.

Microphone A microphone could be used to add narration, voice-overs, or sound effects to the title.

Other Depending on the elements to be incorporated into the multimedia title, other hardware components might include a video camera, a VCR, or a videodisc player.

Figure 2.13
A digital camera

 To extend what you've learned, log on to the Internet at
http://www.thomson.com/wadsworth/shuman
You will find a wide variety of resources and activities related to this chapter.

key terms

- audio card
- binary
- bit
- CD-ROM drive
- development system
- digital camera
- external storage
- gigabyte (GB) *billion*
- kilobyte (KB) *thousand*
- megabyte (MB) *million*
- millisecond (ms)
- monitor
- Multimedia Personal Computer (MPC) specifications
- peripherals
- pixel
- playback system
- processor
- RAM
- resolution
- ROM
- scanner
- upgrade kit
- video card

review questions

1. **T** (**F**) A CD can hold a maximum of 65 MB of information.

2. (**T**) **F** One of the reasons for the recognized quality of Macintosh computers is that Apple controls the specifications of each model.

3. *MPC* stands for _____.

4. The Level 2 MPC specification includes a _____ processor.

5. Screen resolution is measured in the number of dots, called __pixels__, displayed on the screen.

6. (**T**) **F** VGA cards support a screen resolution of 640 × 480.

7. In order to display up to 256 different colors, you need a video card that can support a minimum of __8__ bits.

8. CD-ROM drives are measured in __D.A__ rate and __D.T.__ rate.

9. Circle which of the following bits would be turned on in order to specify a value of 251. (*Note:* This is an 8-bit system.)

    Bits and value   128 64 32 16 8 4 2 1

10. A megabyte is equal to __a million__ characters of data.

projects

1. Complete the following table for each computer system that you work with. Leave blank those areas that do not apply.

	Home Computer	Work Computer	School Computer
Make and model	HP Pavilion 8240		
RAM (in MB)	32 MB		
Processor (type and speed)	233 MHz Pentium Processor		
Hard drive (in MB)	4 GB		
CD-ROM drive (access speed in ms and transfer speed in Kbps)	24x		
Sound (in bits)	32 bit		
Video display (resolution and number of colors)	3D display 1MG V Mem		

Chapter 2 *Hardware Components of a Multimedia System* **43**

2. Using newspaper, magazine, and catalog ads as well as visits to retail stores, choose four computer systems—two playback systems (Macintosh and Windows-based operating systems, or OS) and two development systems—and complete the following.

Playback System:	Macintosh OS	Windows OS
Make and model		
RAM (in MB)		
Processor (type and speed)		
Hard drive (in MB)		
CD-ROM drive (access speed in ms and transfer speed in Kbps)		
Sound (in bits)		
Video display (resolution and number of colors)		
Price		
Source of data		

Development System:	Macintosh OS	Windows OS
Make and model		
RAM (in MB)		
Processor (type and speed)		
Hard drive (in MB)		
CD-ROM drive (access speed in ms and transfer speed in Kbps)		
Sound (in bits)		
Video display (resolution and number of colors)		
Price		
Source of data		

3

Multimedia Elements: Text and Graphics

AFTER COMPLETING THIS CHAPTER YOU WILL BE ABLE TO:

- Describe the text and graphic elements that make up multimedia
- Specify the trade-offs involved in using these elements
- Specify the advantages and disadvantages of using these elements
- Describe the programs used with these elements
- List the primary sources of text and graphics

ARLIER you learned that multimedia titles incorporate text, graphics, sound, animation, and video. In this and the next chapter, you will learn more about these elements, including how they are best used when developing a multimedia title. As you study these, keep in mind that the decision of which elements to use is often a trade-off between cost, time, and effect. And the way in which the elements are used depends on the intended audience and the objectives of the title.

Working with Text

C F R S_c R_c

Perhaps the easiest of all elements to work with is *text*. Most computer users have had experience with word processing and are familiar with entering and editing text and working with fonts and point sizes. Following are some considerations and guidelines to keep in mind when working with text.

BE CONCISE

Reading volumes of text on a computer screen is difficult and tiring. Moreover, it may not be the best way to communicate an idea, concept, or even a fact. The saying "A picture is worth a thousand words" (and perhaps more when sounds, simulations, and animations are added) is as true in multimedia as on the printed page. Although there are certainly titles where text predominates, such as reference works like encyclopedias, combining other elements with text can often reduce the amount of text needed to convey a concept. For example, showing a train pass by a girl standing at a station, and having the train whistle blowing as animated sound waves are displayed, could be an effective way of enhancing the text definition of the Doppler effect. From a design standpoint, text should fill less than half the screen.

USE APPROPRIATE FONTS

Huge, gray blocks of text can be boring to read. You can choose to enliven text by selecting typefaces, called *fonts*, and type sizes appropriate to the audience. Fonts are useful in focusing attention on certain text on the screen, enhancing readability, setting a tone (serious, lighthearted), and projecting an image (progressive, conservative). Figure 3.1 shows two fonts, Critter and ComicsCarToon, that may be appealing to a younger audience because of their childlike or whimsical look. Figure 3.1 also shows the typeface Regency Script, which is more appropriate for a formal look. Fonts can be characterized as serif, sans serif, and decorative. Figure 3.2 shows an example of each of these. A *serif* is a line or curve extending from the ends of a stroke of

Figure 3.1
Examples of fonts

Figure 3.2

Examples of serif, sans serif, and decorative fonts

Bodoni — Serif
ABCDEFGHIJKLMNOPQRSTUVWXYZ

Avant Garde — Sans serif
ABCDEFGHIJKLMNOPQRSTUVWXYZ

Arnold Boecklin — Decorative
ABCDEFGHIJKLMNOPQRSTUVWXYZ

a character. The French word *sans* means without, so a **sans serif** typeface is one without serifs.

When choosing a font, always consider the objectives and the audience. All decisions in developing a multimedia title ultimately depend on the objectives of the title and the intended audience. If the objectives have to do with creating a reference title such as the "Selected Works of Shakespeare" and the audience is college students, the title would be text intensive and parts of the text, such as headings, might utilize a decorative font appropriate to Shakespeare's time.

MAKE IT READABLE

The overriding concern with text is readability. Although a decorative font may be attractive, it may also be hard to read. And though it may seem important to include a great deal of text, filling the screen with text or reducing the size of the type to accommodate more text might also hinder readability. Sans serif text is clean, simple, and projects rationality and objectivity (although not always readability). Serifs create the illusion of a line along the base of a line of text and guide the eye across the screen, facilitating readability. Sans serif type does not have such a guide, and the eye can have difficulty remaining focused on a line of text, being inclined instead to leave the text line and wander through the body of the text. For body type, a serif font is preferred for readability. Research has shown that comprehension of

Figure 3.3

Text set in serif and sans serif fonts. Which is easier to read?

Serif type	Sans serif
Fourscore and seven years ago our fathers brought forth on this continent a new nation, conceived in Liberty, and dedicated to the proposition that all men are created equal. Now we are engaged in a great civil war, testing whether that nation, or any nation so conceived and so dedicated, can long endure. We are met on a great battlefield of that war. We have come to dedicate a portion of that field, as a final resting-place for those who here gave their lives that that nation might live. It is altogether fitting and proper that we should do this.	Fourscore and seven years ago our fathers brought forth on this continent a new nation, conceived in Liberty, and dedicated to the proposition that all men are created equal. Now we are engaged in a great civil war, testing whether that nation, or any nation so conceived and so dedicated, can long endure. We are met on a great battlefield of that war. We have come to dedicate a portion of that field, as a final resting-place for those who here gave their lives that that nation might live. It is altogether fitting and proper that we should do this.

text blocks with serifs is 75 to 80 percent, whereas comprehension of text blocks set in sans serif typefaces is 20 to 30 percent. Serif text is described as old-fashioned, friendly, and easy to read. Sans serif text is described as clean, sleek, modern, and not so easy to read. Figure 3.3 shows text set in serif and sans serif fonts to illustrate this point. Notice in the sans serif text how each stroke of every letter has the same width. Contrast this with serif text, where each letter has thick and thin strokes. This contributes to a serif font's readability. A sans serif font may be used for headings, providing contrast with a serif font for text. Sans serif fonts may be used for multimedia titles in which there is not much text, such as a game. Decorative fonts are best used for emphasis.

Fonts are measured in ***point sizes***. There are 72 points per inch. Ten and 12 point are common sizes for type displayed on the screen. The size often depends on the application. For example, text that appears as a title at the top of a screen may be relatively large, whereas text that is used on a button might be quite small. Suggested guidelines are as follows.

Headings	14 to 48 point
Subheadings	Half the heading size
Text blocks	10 to 12 point

Headings and subheadings are used to attract attention and provide the user with quick identification of the screen contents, while text blocks provide the substance.

CONSIDER TYPE STYLES AND COLORS

Three common *type styles* are **bold**, *italic*, and underline. These styles are often used for emphasis in print materials. In multimedia applications, however, they are more often used to indicate hypertext, or hotwords. Clicking on **hypertext** will display additional text (such as a definition), or cause some action (such as a sound or animation), or jump to another part of the application. Figure 3.4 shows an example of the use of italic type to suggest that the word *hotwords* can be clicked on to cause an action, in this case displaying a pop-up window with a definition.

USE RESTRAINT AND BE CONSISTENT

Although it may be tempting and certainly easy to use various typefaces, sizes, and styles, it is important to exercise restraint. Be careful to avoid the "ransom note" effect: a busy and difficult-to-read design resulting from too many fonts and type styles on one screen. In addition, you should try to maintain consistency in the use of text. For example, if several screens have

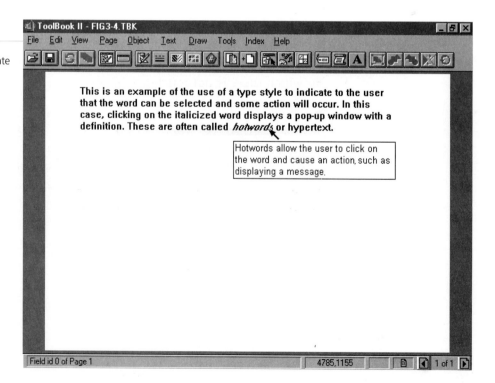

Figure 3.4

Using a type style to indicate a hotword

interactive exercise

On the *Multimedia in Action* CD are examples of text with different fonts, point sizes, and type styles. Viewing these examples will help you understand the use of fonts.

1. Start the CD.
2. Choose Text from the contents screen and read the instructions.
3. Click on the questions button and review the questions for Fonts, Font Sizes, and Text Styles.
4. Click on the demos button and choose Fonts, Font Sizes, and Text Styles.
5. After viewing the demo, respond to the questions previously reviewed.

a similar heading, you should use the same typeface, size, and style for all such headings.

An important consideration in selecting fonts is whether or not they are available on the playback system. Both Macintosh and Windows-based computers have built-in fonts. If your title uses a font that is not on these platforms, the playback system will try to match your font with as close a substitute as possible. This can have disastrous effects on the appearance of the text by causing changes in word spacing, wrapping the text inappropriately, and even altering the size of the type. To avoid these problems, you can bundle the font with your title so that it is always available. PostScript fonts for the Macintosh as well as TrueType fonts, such as Helvetica and Times on the Macintosh and Arial and Times New Roman on the Windows-based machines, are usually installed with the operating system and thus available on most of these computers. Another way to avoid font compatibility problems is to create graphic images, or bitmaps, of text. Graphic programs allow you to type in text and then save the text block as a graphic image. The graphic can then be placed in a multimedia title without regard to available fonts, appearing to the user as typed text. One drawback of this process is that the graphic image takes up more memory than the typed text. Another drawback is that the text cannot be easily edited.

Accommodating Text-Intensive Titles

There are times when a title must include a great deal of text. Reference titles such as encyclopedias are good examples. There are two ways to accommodate large amounts of text without overwhelming—and perhaps turning off—the user. First, when possible, use other ways to communicate the message. Ask yourself: Can the idea or content be communicated in some other way? For example, show an animation or use narration rather than write about the idea. Second, consider using a small amount of text and then allowing the user to obtain more information as desired, using one of the following techniques.

HYPERLINKING

Allow the user to select a hotword or a graphic or button to jump to another part of the title that displays more text.

POP-UP MESSAGES, SCROLL BARS, AND DROP-DOWN BOXES

Figure 3.4 shows a pop-up message that is displayed when the user clicks on a hotword. Figure 3.5 shows a scroll box that displays more information as the user clicks on the down arrow or drags the button on the scroll bar. Figure 3.5 also shows a drop-down box that can display information when the user clicks on the down arrow. A drop-down box is often used to display a menu of choices from which the user can select.

Figure 3.5

A scroll box and a drop-down box.

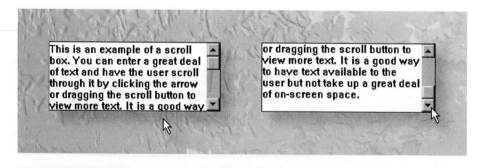
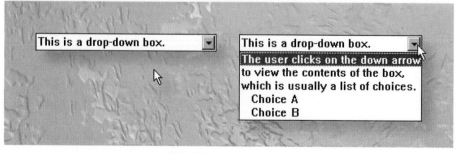

interactive exercise

On the *Multimedia in Action* CD are examples of the ways to deal with large amounts of text. Take a few minutes to view these.

1. Start the CD.
2. Choose Text from the contents screen and read the instructions.
3. Click on the questions button and review the questions for Techniques for Text-Intensive Titles.
4. Click on the demos button and choose Techniques for Text-Intensive Titles.
5. After viewing the demo, respond to the questions previously reviewed.

Software for Creating and Editing Text

Word processing programs, such as Microsoft Word and WordPerfect, are useful in creating text for titles that are text intensive. Once text is created in a word processing program, it can easily be copied to a multimedia title. If the title is not text intensive, it may be more efficient to use graphics programs (software used to draw and paint images, such as CorelDRAW and Adobe Illustrator) and authoring programs (software used to create multimedia titles) to generate the text. These programs have text tools that allow you to enter and edit text and select fonts, point sizes, and type styles and colors. They also allow you to create special effects, such as distorting or animating the text. Font packages can be purchased that provide a variety of specialized fonts, and programs are available that allow you to create your own fonts. Other sources of text include text that is already in electronic form and text that can be scanned.

If you were creating a CD of the Yellow Pages directory for a particular geographical area, for example, you would not retype all of the text from the current directory. Rather, you would obtain the electronic files already used to publish the current directory and simply import or copy the data into your title. If a particular printed document was not available in electronic form, or you needed only small parts of it, you could use a scanner and an ***optical character recognition (OCR)*** program to capture the desired text. As the document is scanned, the OCR program translates the text into a format that can be used by a word processing program.

Working with Graphics

The introduction of the Apple Macintosh computer and the Microsoft Windows program changed the way we worked with computers. Using a mouse and a desktop metaphor, we click on icons and drop-down menus, drag folders, and resize windows. We are accustomed to working with graphical images on the screen and, in fact, expect to see them. Graphics such as drawings and photographs are integral to multimedia titles. Visualization is an important part of the communications process, and graphical images can be used to add emphasis, direct attention, illustrate concepts, and provide a background for the content.

There are two categories of graphics: draw-type and bitmaps. **Draw-type graphics**, also called vector graphics, represent an image as a geometric shape made up of straight lines, ovals, and arcs. When a line is drawn, a set of instructions is written to describe its size, position, and shape. If more than one line is drawn, it has a precise relationship to the other parts. Figure 3.6a shows a graphic, a pie chart, made up of a circle and lines. The instructions that create the circle and lines establish the relationship between them. If a change is made, say, in the size of the circle, the relationship between the circle and the lines stays the same. Figure 3.6b shows the graphic reduced in size and rotated. The reduced graphic keeps

Figure 3.6

A draw-type graphic made up of a circle and lines

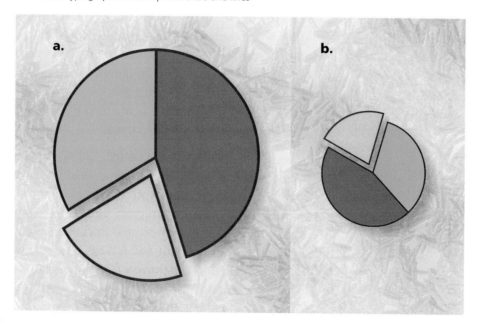

the same relationship (relative position and relative size) as the original graphic. The ability to resize and rotate a graphic without distortion is a major advantage of draw-type graphics. Another advantage is their smaller file size. Because these graphics are stored as sets of instructions, the file sizes can be significantly smaller than bitmaps. One of the drawbacks of draw-type graphics is that the more complex they are, the larger the file size and the longer they take to appear on the screen. Another disadvantage is that they cannot display photorealistic quality.

A ***bitmap*** represents the image as an array of dots, called ***pixels***. As you learned in chapter 2, the screen is made up of a grid, and each part of the grid is a picture element (pixel). Color information, called color depth, is recorded for each pixel. Depending on the number of colors used, a bitmap file can be relatively small.

GRAPHIC IMAGE QUALITY

Because draw-type graphics are displayed using a set of instructions that define a line, they are not as discrete as bitmaps. The quality of the image is therefore lower. Creating a circle with a draw-type program allows you to specify only one color for the entire circle; but creating a bitmap circle allows you to change the color of every pixel in the circle. Thus the bitmap can be more photorealistic. The trade-off is that bitmap graphic files are larger than draw-type files. File size is a function of the image size and the color depth.

IMAGE SIZE, COLOR DEPTH, AND FILE SIZE

In chapter 2 you learned that screen resolution is measured in horizontal and vertical pixels, with 640 × 480 being standard. The more pixels per inch on the screen, the finer the detail and hence the better the image quality. A screen resolution of 1024 × 768 displays a much better quality image than 640 × 480 on the same size monitor. You also learned that various numbers of colors can be associated with each pixel, depending on the number of bits specified (8-bit, 256 colors; 16-bit, 65,536 colors; 24-bit, 16.7 million colors). The range of colors available for pixels is called the ***color depth***. The file size of a bitmap graphic is related to its image size and color depth. It can be estimated using the following formula:

image size in pixels × color depth in bits ÷ 8

Figure 3.7

The file size given the image size and color depth of bitmap graphics

Image Size in Pixels	Screen Size	Color Depth in Bits	Number of Available Colors	File Size in Bytes (approximate)
640 × 480	Full screen	8	256	300,000
320 × 240	Quarter screen	8	256	77,000
1024 × 768	Full screen	24	16.7 million	2,400,000

You divide by 8 because file size is measured in bytes and there are 8 bits per byte. Figure 3.7 shows examples of file sizes given the image size and color depth. Figure 3.8 shows the image quality of various bitmap graphics. The minimum requirements for a Level 3 MPC system is 640 × 480 resolution and 65,536 colors. Therefore, creating a multimedia title using 1024 × 768 and 24 million colors would significantly reduce the number of potential users.

Figure 3.8

The image quality of various bitmap graphics

Software for Creating and Editing Graphics

Graphics programs are the tools that allow an artist to create and edit designs used in multimedia titles. There are dozens of graphics programs; some come with operating systems, such as Microsoft Paint which comes with Windows 95, and others are included in authoring programs used to create multimedia titles (see chapter 5). These are relatively unsophisticated programs, however, lacking many features found in high-end applications. Generally, graphics programs can be categorized as drawing, paint, and image-editing programs.

Drawing programs—those creating draw-type graphics—provide for freehand as well as geometric shapes and are useful in creating designs where precise dimensions and relationships are important. Figure 3.9 shows an example of a drawing program, Adobe Illustrator.

Paint programs—those creating bitmaps—are useful in creating original art, because they provide the tools (brushes, pens, spray paint) used by artists. Figure 3.10 shows an example of a paint program, Paint Shop Pro.

Image-editing programs are useful for making changes to existing images, such as manipulating the brightness or contrast, or applying textures or

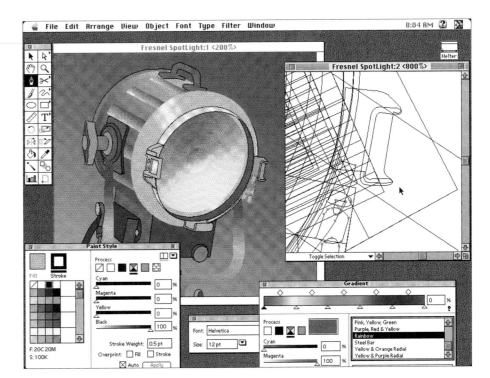

Figure 3.9
A drawing program

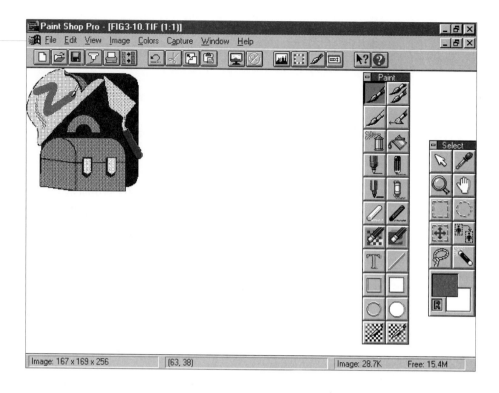

Figure 3.10
A paint program

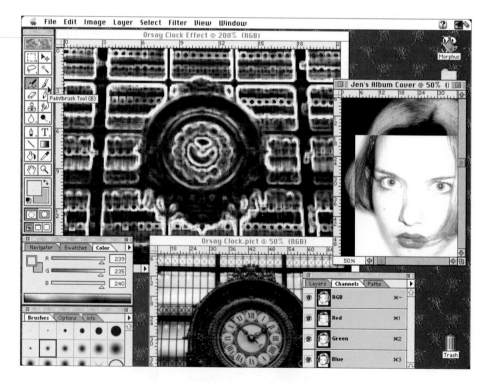

Figure 3.11
An image-editing program

patterns. Figure 3.11 shows an example of an image-editing program, Photoshop. Several of the high-end programs have features from all three of these categories—drawing, paint, and image-editing.

Features of Graphics Programs

Following is a list of features that are available on high-end graphics programs. Examples of several of these features are found on the *Multimedia in Action* CD. You will be referred to the examples as you read this section.

interactive exercise

On the *Multimedia in Action* CD are examples of images created using various features of graphics programs.

1. Start the CD.
2. Choose Graphics from the contents screen and read the instructions.
3. Click on the questions button and review the questions.
4. Click on the demos button and choose Features.
5. After viewing the demo, respond to the questions previously reviewed.

Type of graphics program The program is primarily a drawing, a paint (bitmap), or an image-editing program. Many programs allow you to create both draw- and paint-type graphics.

Cross-platform compatibility The program comes in both a Windows and a Macintosh version, and/or is able to create graphics that can be used on both the Macintosh and Windows platforms.

Graphics file support The program allows saving and/or converting graphic images using several of the more popular file formats, such as TIFF (Tagged Image File Format), BMP (bitmap), PCX (Windows Paint), PICT (Macintosh picture format), and so forth. With the enormous interest in the World Wide Web, JPEG (Joint Photographic Experts Group) and GIF (Graphics Interchange Format—the CompuServe file format) files are becoming more important, because they are the standard file formats for the World Wide Web.

Layers The program supports object layering, which allows you to include more than one bitmap in an image and edit each bitmap independently of the others.

Image enhancement with painting tools The program has pencil, brush, airbrush, text, and line tools; user-defined brushes and the ability to preview the brush size; and an option to paint with textures and patterns and to retouch using smudge, sharpen, and blur features.

Choose Image Enhancement and view the graphics.

Selection tools The program allows selection of any part of an image using a freehand tool, including selection of all the pixels of a certain color. It also allows the use of masks to isolate parts of an image and apply a special effect such as a drop shadow.

Color adjustments The program allows you to adjust image color and choose from a range of colors simultaneously. You can selectively change hue (the shade or color itself), saturation (the relative brillance or vibrancy of a color), and brightness.

Choose Color Adjustment and view the graphics.

Image manipulation The program can stretch, skew, and rotate an image.

Choose Image Manipulation and view the graphics.

Filters The program has filters for sharpening, softening, and stylizing the image.

Choose Image Filters and view the graphics.

Antialiasing The program supports antialiasing. Because bitmaps are made up of rectangular pixels, the outside edge of the image can appear jagged as shown in figure 3.12. *Antialiasing* smoothes the edges by blending the colors on the edge of the image with the adjacent colors as shown in figure 3.13.

Choose Image Antialiasing and view the graphics.

Text support The program allows manipulation of PostScript and TrueType fonts (standard font types for the Macintosh and Windows operating systems).

Graphics tablets The program supports pressure-sensitive graphics tablets as shown in figure 3.14.

Figure 3.12
Jagged edges of a bitmap

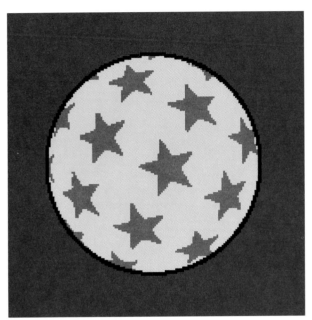

Figure 3.13
Antialiasing smoothes out the jagged edges of the bitmap

Figure 3.14
A pressure-sensitive graphics tablet

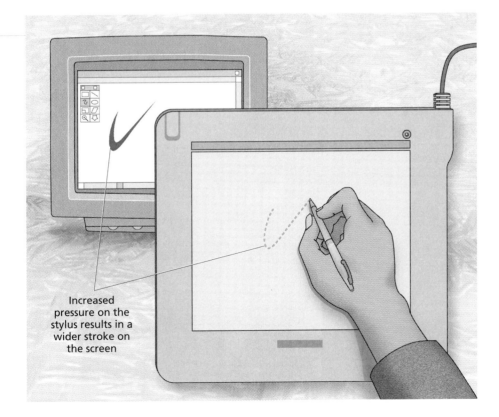

Increased pressure on the stylus results in a wider stroke on the screen

Open architecture   The program is compatible with third-party software, such as programs that provide special effects.

Sources of Graphic Images

Draw- and paint-type programs are used to develop graphic images from scratch. Following are other ways to generate graphics that can be incorporated into a multimedia title.

Clip art, stock photographs, and fine art   Some graphics programs come with clip art and stock photographs, but these are often limited or of poor quality. Several companies specialize in providing images for multimedia titles. For example, Photodisc, a leader in digital stock images, provides more than 50,000 photographs that can be purchased on CD (see figure 3.15) or

Figure 3.15

Clip art and stock photographs

accessed from its Web site. Corel Gallery 2 has 15,000 clip-art images, and Corel Stock Photo Library has 20,000 photographs.

These large CD libraries have a function that allows you to search for an image using descriptive words. For example, if you wanted to view photos of sunsets, you might search using the word *sun* or *sunset*, and a list of photos would appear. The companies that sell clip art and stock photographs often allow you to use the graphics in a multimedia title and to distribute the title without paying a royalty. The Corbis Corporation provides images from a wide variety of sources, including leading professional photographers, museums, cultural institutions, public and private archives, and private collections. Corbis has more than 17 million images, including the Ansel Adams collection. Chapter 12 discusses the copyright issues related to using graphic images in multimedia titles.

Video images Pictures from video sources such as VCRs, videodiscs, and video cameras can be transferred to a computer, and individual frames can become graphic images.

Still images Digital cameras specifically designed to capture images in a digital form are useful in generating graphics. You take a picture as you would with any still camera. Then the camera is connected to a computer, and the images (pictures) are transferred from the camera to the computer. The programs that come with the camera allow you name, sort, resize, rotate, and save the images. Photographs that are taken with a regular camera can be made into a graphic image by scanning directly into a computer or writing to a Photo CD. (See chapter 9 for a discussion of Photo CDs.) Photo CDs can hold approximately 100 images, and many film-processing outlets can create a Photo CD from an ordinary roll of film.

Scanned images One of the most useful pieces of equipment for generating graphics is a ***scanner***. Figure 3.16 shows a scanner and scanning program used to digitize essentially any image that can be placed in the scanner. Depending on the type of scanner and the sophistication of the program, you can zoom in and crop (select) specific parts of the image before capturing it. You can edit the image by adjusting colors, contrast, and brightness. All sorts of images can be scanned, including photographs, illustrations from books, drawings, slides (using a slide scanner)—even objects, although they will display in two-dimensional form. Using a scanner to capture images from objects such as leaves, bricks, fabric, or

Figure 3.16

A scanner and scanning program used to digitize images such as photos and illustrations

aluminum foil provides a way to generate creative graphics that can be used for backgrounds in a multimedia title (see figure 3.17).

Screen capture programs Both Macintosh and Windows-based computers allow you to capture as a graphic whatever is on the computer screen. When you hold down the Command and Shift keys and press 3 on the Macintosh, whatever is on the screen is captured as a graphics file. When you press the Print Screen key on a Windows-based computer, whatever is on the screen is placed on the Clipboard and can be pasted into a graphics program and then saved as a graphics file. In addition, there are programs such as Hijack Pro and Collage Plus that are specifically designed to capture a screen or part of a screen and save it as a graphics file of a type you specify. Capturing a screen can be useful if you are working with text that you want to display as a graphic. You could type the text using a word processing program and then capture it on-screen as a graphic. Text displayed as a graphic helps ensure that it will appear the same on different playback systems because fonts are not involved.

To extend what you've learned, log on to the Internet at

http://www.thomson.com/wadsworth/shuman

You will find a wide variety of resources and activities related to this chapter.

Figure 3.17

Examples of scanned objects that can be used as backgrounds

key terms

antialiasing
bitmap
color depth
drawing programs
draw-type graphic
font
graphics programs
hypertext
image-editing programs

optical character recognition (OCR)
paint programs
pixel
point size
sans serif
scanner
serif
text
type style

review questions

1. Guidelines to keep in mind when using text in a multimedia title are:

   a. C_____ Re_____
   b. F_____
   c. R_____
   d. Se_____

2. (T) F   A serif font provides better readability for body text.

3. (T) F   Some graphics programs allow you to save text as a graphic image.

4. (T) F   Hyperlinking is a way to accommodate text-intensive titles.

5. The two categories of graphics are:

   a. DRAW-TYPE (VECTOR)
   b. BIT MAPS

6. T (F)   A draw-type graphic represents the image as an array of dots, called pixels.

7. (T) F   A bitmap graphic can be more photorealistic than a draw-type graphic.

8. T (F)   Bitmap graphics are smaller in size than draw-type graphics.

9. The file size of an image is a function of:
   a. _IMAGE SIZE (PIXELS)_
   b. _COLOR DEPT BITS_

10. What is the approximate file size in kilobytes of a 16-bit color image that displays at a resolution of 640 × 480 pixels?

$$\frac{640 \times 480 \times \overset{2}{16}}{8}$$

projects

1. Research two popular graphics programs (such as Illustrator, Freehand, or CorelDRAW) and prepare a report that compares them based on the features of graphics programs discussed in this chapter. Include the prices for each program.

   Prepare a brief oral presentation of your report and be ready to present it to your class.

2. Study two multimedia titles, a children's title and a reference title (such as an encyclopedia). Prepare a report comparing the titles in terms of their use of the following:

   - Fonts (type styles, sizes, serifs)
   - Text (too much, too little, techniques used to accommodate large amounts of text)
   - Graphics (difference in the number of colors used, the quality of the graphics)

   Prepare a brief oral presentation of your report and be ready to present it to your class.

4

Multimedia Elements: Sound, Animation, and Video

AFTER COMPLETING THIS CHAPTER YOU WILL BE ABLE TO:

- Describe the sound, animation, and video elements that make up multimedia
- Specify the trade-offs involved in using these elements
- Specify the advantages and disadvantages of using these elements
- Describe the programs used with these elements

IN chapter 3 you learned how text and graphics are used in multimedia titles. In this chapter you will learn about the other elements: sound, animation, and video. These elements are relatively new to desktop computers and can greatly enhance the effectiveness of multimedia titles. As you study these, keep in mind that the decision of which elements to use is often a trade-off between cost, time, and effect. And the way in which the elements are used depends on the intended audience and the objectives of the title.

Sound

In the early days of using desktop computers, usually the only sound that you heard was a beep—often accompanied by an error message. Now a whole range of sounds can be played through a computer, including music, narration, sound effects, and original recordings of such events as a presidential address or a rock concert. Sounds are critical to multimedia. Often sound provides the only effective way to convey an idea, elicit an emotion, or dramatize a point. How would you describe in words or show in an animation the sound a whale makes? Think about how chilling it is to hear the footsteps on the stairway of the haunted house. Consider how useful it is to hear the pronunciation of *"Buenos dias"* as you are studying Spanish.

To study the use of sound in computers, you need a basic understanding of sound. When we speak, vibrations, called sound waves, are created. These sound waves have a recurring pattern, shown in figure 4.1, that is called an **analog wave pattern**. The wave pattern has two attributes that affect how you work with sound on a computer: volume and frequency. The height of each peak in the sound wave relates to its **volume**—the higher the peak, the louder the sound. The distance between the peaks is the **frequency**—the greater the distance, the lower the pitch. Frequency is measured in **hertz (Hz)**. A pattern that recurs every second is equal to 1 Hz. If the pattern recurs 1,000 times in 1 second, it would be 1,000 Hz, or 1 kHz (kilohertz).

Figure 4.1

An analog wave pattern

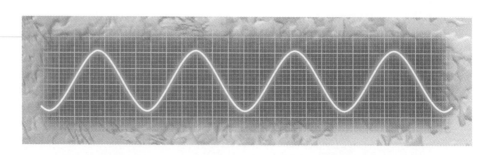

Figure 4.2

An analog wave pattern and a digital sampling of the pattern

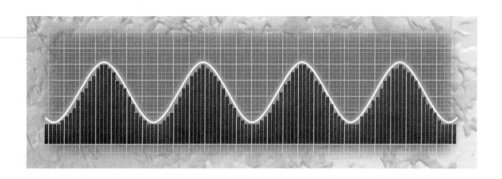

SAMPLING

In order for a computer to work with sound waves, they must be converted from analog to digital form. This is done through a process called *sampling*, in which every fraction of a second a sample of the sound is recorded in digital bits. There are two factors that affect the quality of the digitized sound: the number of times the sample is taken, called the *sample rate;* and the amount of information stored about the sample, called the *sample size*. Figure 4.2 shows an example of an analog wave pattern that has been sampled.

The three most common sample rates are 11.025 kHz, 22.05 kHz, and 44.1 kHz. The higher the sample rate, the more samples that are taken and, thus, the better the quality of the digitized sound. The two most common sample sizes are 8 bit and 16 bit. An 8-bit sample allows 256 values that are used to describe the sound, whereas a 16-bit sample provides 65,536 values. Again, the greater the sample size, the better the quality of the sound. Figure 4.3

Figure 4.3

File size of 10 seconds of digitized audio in stereo

Sample Rate	Bit Value	Size of File
44.1 kHz	16	1.76 MB → CD QUALITY SOUND
44.1 kHz	8	882 KB
22.05 kHz	16	882 KB
22.05 kHz	8	440 KB
11.025 kHz	8	220 KB → TELEPHONE

Figure 4.4

A sound card used to digitize sound

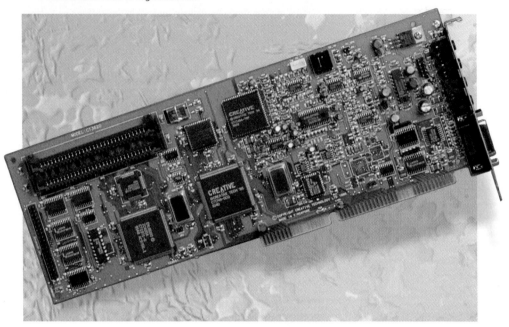

shows the file size (in bytes) for 10 seconds of digital audio given various sample rates and bit values. The following formula is used to determine the byte size of a 1-second recording:

> sample rate × bit value ÷ 8

Thus a 1-second recording at a sample rate of 44.1 kHz and 16 bits would be 88,200 bytes (44,100 × 16 ÷ 8). You would double this number for a stereo recording. The 44.1 kHz 16-bit sample is CD quality, whereas the 11.025 kHz 8-bit sample would be marginal quality.

A *sound card*, shown in figure 4.4, is used to digitize sound. A sound from some external source is sent to the card. The external source could be a cassette tape, videotape player, CD, or someone speaking into a microphone. The sound card samples the sound based on the sample rate and bit value and produces the digital approximation of the analog signal.

Once a sound has been digitized, it can easily be manipulated using a sound-editing program. Figure 4.5 shows a sound-editing program and some of its features. Using the mouse pointer, you can select part of the recording and cut it out of the pattern or replace it with another sound. Or you can choose to add sound effects such as an echo or fade-in and fade-out.

Chapter 4 *Multimedia Elements: Sound, Animation, and Video* **73**

Figure 4.5

A sound-editing program

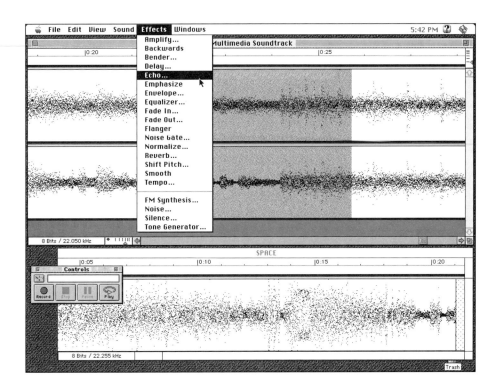

interactive exercise

On the *Multimedia in Action* CD are sound clips recorded at different sample rates and bit values. Take a few moments to play these clips and notice the differences in the sounds relative to their sample rate and bit value. Also on the CD are clips of various sound effects. Take a few moments to listen to these to get an idea of how sounds can be edited to create different sound effects.

1. Start the CD.
2. Choose Sound from the contents screen and read the instructions.
3. Click on the questions button and review the questions for the Sampling option.
4. Click on the demos button and choose Sampling.
5. After viewing the demo, respond to the questions previously reviewed.
6. Click on the questions button and review the questions for the Sound Effects option.
7. Click on the demos button and choose Sound Effects.
8. After viewing the demo, respond to the questions previously reviewed.

MIDI

Another way to incorporate sound into multimedia titles is with **MIDI** files. **MIDI**, which stands for **Musical Instrument Digital Interface**, is a standard format that enables computers and electronic musical instruments to communicate sound information. Digitized audio begins with a sound, samples it, and creates a digital representation which is stored in a file. MIDI begins with an *event*, such as pressing a key on an electronic musical instrument such as a MIDI keyboard, and codes the event (including which key and how hard and long it was pressed) as a series of commands. These are stored in a file and can be sent from the computer to another instrument, such as a synthesizer, for playback.

MIDI has been compared to a musical score, because it represents the notes that are played, along with such information as volume and frequency, rather than the sound itself. This level of detail allows more-precise editing than with digitized sound. Also, because MIDI files contain code instead of the actual digitized sounds, they can be hundreds of times smaller than audio files. On the other hand, working with MIDI requires specialized software and may require additional equipment for recording and playback, or a MIDI-compatible audio card.

Animation

We often think of *animation* as full-length Disney movies and Saturday-morning cartoons in which illustrated heroes and villains and especially animal characters come to life. Television programs, movies, and videos are part of our daily lives. Animation plays a huge role in entertainment (providing action and realism) and education (providing visualization and demonstration). Entertainment multimedia titles in general, and children's titles specifically, rely heavily on animation. But animation can also be extremely effective in other titles, such as training applications. For example, say a mechanic needs to be trained on a hydraulic system for the landing gear of a jet plane. The training includes information on the flow of hydraulic fluid through the system. It might be impossible to videotape the actual flow of the fluid inside the landing gear, but an animation could provide a simulation and even dramatize how pressure is created during the process.

The perception of motion in an animation is an illusion. The movement that we see is, like a movie, made up of many still images, each in its own frame. Movies on video run at about 30 *frames per second (fps)*, but computer animations can be effective at 12 to 15 fps. Anything less results in a jerky motion, as the eye detects the changes from one frame to the next.

2-D ANIMATION

There are two types of 2-D animation, cel and path. **Cel animation** is based on the changes that occur from one frame to another to give the illusion of movement. *Cel* comes from the word *celluloid* (a transparent sheet material) which was first used to draw the images and place them on a stationary background. Figure 4.6 shows an example of cel animation. Notice that the background remains fixed as the object changes from frame to frame. You could have more than one object move against a fixed background.

Figure 4.6
An example of cel animation

Computer-based cel animation is usually done with animation programs, although some multimedia authoring programs can create cel animations.

Path animation moves an object along a predetermined path on the screen. The path could be a straight line or it could include any number of curves. Often the object does not change, although it might be resized or reshaped. Figure 4.7 shows path animation used to create the illusion of a bouncing ball. This can be a relatively easy process, because you need only one object (the ball), rather than several objects as in figure 4.6. Path animations can often be created using a multimedia authoring program by simply dragging the mouse pointer around the screen, or by pointing to different locations on the screen and clicking the mouse button. Some authoring programs even allow you to set the object's beginning position on one frame and its ending position on another frame; then the program uses a technique called ***tweening*** to automatically fill in the intervening frames.

Figure 4.7

An example of path animation

interactive exercise

On the *Multimedia in Action* CD are examples of both cel and path animations. Take a few minutes to view these.

1. Start the CD.
2. Choose Animation from the contents screen and read the instructions.
3. Click on the questions button and review the questions for 2-D animation.
4. Click on the demos button and choose 2-D animation.
5. After viewing the demo, respond to the questions previously reviewed.

3-D ANIMATION

Although 2-D animation can be effective in enhancing a multimedia title, 3-D animation takes the entire experience of multimedia to another level. Three-dimensional animation is the foundation upon which many multimedia CD games and adventure titles are constructed. Top-selling products such as Myst and 7th Guest use 3-D animation to bring the user into the setting and make him or her seem a part of the action. Whether opening doors, climbing stairs, or exploring mysterious rooms, the user is a participant, not a spectator. Creating 3-D animation is considerably more complex than 2-D animation and involves three steps: modeling, animation, and rendering.

Modeling is the process of creating 3-D objects and scenes. One technique, shown in figure 4.8, involves drawing various views of an object (top, side, cross-section) by setting points on a grid. These views are used to define the object's shape. The animation step involves defining the object's motion and how the lighting and views change during the animation. **Rendering** is the final step in creating 3-D animation and involves giving objects attributes such as colors, surface textures, and degrees of transparency (see figure 4.9). Rendering can take considerable time (days), depending on the complexity of the animation. There are different rendering processes, varying in time needed and quality of the completed animation. Animators therefore often render the animation using a quicker, lower-resolution process as a test. Then they use a slower, higher-quality process for the finished animation. Strata Pro 3D, Swivel 3D, and 3D Studio are examples of programs that can produce quite sophisticated three-dimensional animations.

Figure 4.8
3-D modeling

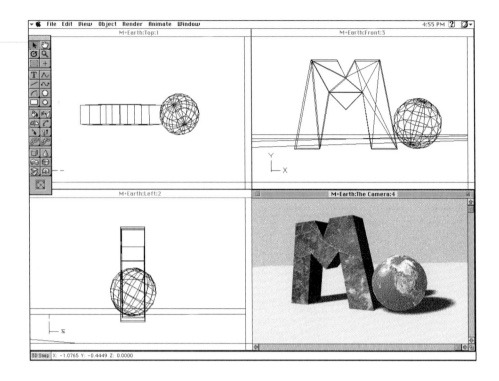

Figure 4.9
Rendering a 3-D animation

Chapter 4 *Multimedia Elements: Sound, Animation, and Video*

interactive exercise

On the *Multimedia in Action* CD are examples of 3-D animations. Take a few minutes to view these.

1. Start the CD.
2. Choose Animation from the contents screen and read the instructions.
3. Click on the questions button and review the questions for 3-D animation.
4. Click on the demos button and choose 3-D animation.
5. After viewing the demo, respond to the questions previously reviewed.

ANIMATION SPECIAL EFFECTS

A common special effect for animations is morphing. **Morphing** is the process of transforming one image into another via a series of frames as shown in figure 4.10. An example would be a photo kiosk in which two people would have their pictures taken and a photo would be generated by the computer

Figure 4.10
Morphing

Figure 4.11
Warping

that combines their images. The morphing process involves selecting sets of corresponding points on each of the images. Thus, in morphing the two faces shown in figure 4.10, the sets of points might include eyes, lips, ears, and outline of the head. Based on these sets of points, the morphing program rearranges the pixels to transition the original image into the second via a series of intervening images. Morphing is useful not only in showing how two images blend together, but also in showing how an image might change over time. **Warping** is a related special effect that allows you to manipulate a single image. For example, you could stretch a facial feature to change a frown into a smile (see figure 4.11).

interactive exercise

On the *Multimedia in Action* CD are examples of morphing and warping. Take a few minutes to view these.

1. Start the CD.
2. Choose Animation from the contents screen and read the instructions.
3. Click on the questions button and review the questions for Morphing and Warping.
4. Click on the demos button and choose Morphing and Warping.
5. After viewing the demo, respond to the questions previously reviewed.

Virtual Reality

Virtual reality (VR) attempts to create an environment that surrounds the user so that he or she becomes part of the experience. The term *virtual reality* has been used to describe various types of applications, some that are more experiential than others, as the following examples illustrate.

- Boeing uses flight simulators that are cockpits of actual airplanes, mounted on hydraulically controlled legs that can simulate every motion of an aircraft. Flight crews training in these simulators can be presented with any number of environments (airports, weather conditions, landing approaches) on displays viewed through the cockpit windows. These simulators are so realistic that the Federal Aviation Administration (FAA) will approve current pilots for certification on a specific model using the simulator alone.

- In some compact disc–based adventure games, the surroundings change as the user points the mouse cursor and walks through doors, up stairs, turns left or right, or otherwise moves through the various scenes. The goal of the multimedia developer is to make it seem as though the user is actually standing in the middle of a room, or in an arcade, or in a haunted house, and so on. Related applications are those that allow a walk-through of a building. A popular title is a walk-through of the White House, which allows the user to view different rooms and to zoom in on objects such as paintings and sculptures.

- There are arcade-type games and educational applications that require headgear with goggles that allow the user to "step into" a virtual world. As the user turns his head, a different view of the world appears. Gloves and handheld equipment can be used to allow the user to interact with the environment.

Although most virtual reality applications are animations, Apple Computer has developed a QuickTime VR system. This system starts with photographs taken in a panoramic format. This is accomplished by mounting a digital camera on a tripod that allows the user to take a series of still pictures. The camera is rotated a few degrees after each picture, until a complete 360-degree panorama is obtained. These photographs are electronically "stitched" together to provide a seamless 360-degree view. The files created with this system can be brought into a multimedia title that allows the user to point the mouse cursor to navigate around the setting and to zoom in on any object.

Because virtual reality is so 3-D graphic intensive, it is not as applicable to home and school CD titles that are played on typical multimedia computers.

interactive exercise

On the *Multimedia in Action* CD is an example of a virtual reality title. Take a few minutes to view this example.

1. Start the CD.
2. Choose Animation from the contents screen and read the instructions.
3. Click on the questions button and review the questions for Virtual Reality.
4. Click on the demos button and choose Virtual Reality.
5. After viewing the demo, respond to the questions previously reviewed.

Video

The ability to incorporate digitized video into a multimedia title marked an important achievement in the evolution of the multimedia industry. Consider the following: You are developing a report on the civil rights movement in the United States to be presented to your fellow students. The purpose is to inform them of various significant events. You want to include excerpts from Martin Luther King Jr.'s "I Have a Dream" speech. You could:

- Type part of the speech and hand it out (text)
- Show a photo of Martin Luther King Jr. (graphics)
- Play an audio excerpt of the speech (sound)
- Play a video excerpt of the speech (video)

Those viewing the video would recognize the impact of seeing the actual event rather than simply reading about it or listening to it. Video brings a sense of realism to multimedia titles and is useful in engaging the user and evoking emotion.

DIGITIZING THE VIDEO SIGNAL

Video, like sound, is usually recorded and played as an analog signal. It must therefore be digitized in order to be incorporated into a multimedia title. Figure 4.12 shows the process for digitizing an analog video signal. A video source, such as a video camera, VCR, TV, or videodisc, is connected to a video capture card in a computer. As the video source is played, the analog signal is sent to the video card and converted into a digital file that

Figure 4.12

The process for digitizing an analog video signal

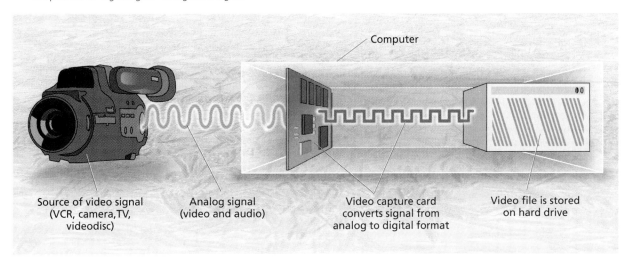

is stored on the hard drive. At the same time, the sound from the video source is also digitized.

One of the advantages of digitized video is that it can be easily edited. Analog video, such as a videotape, is linear; there is a beginning, middle, and end. If you want to edit it, you need to continually rewind, pause, and fast-forward the tape to display the desired frames. Digitized video, on the other hand, allows random access to any part of the video, and editing can be as easy as the cut-and-paste process in a word processing program. In addition, adding special effects such as fly-in titles and transitions is relatively simple.

FILE SIZE CONSIDERATIONS

Although digitized video has many advantages, there is an important consideration: file size. Digitized video files can be extremely large. A single second of high-quality color video that takes up only one-quarter of a computer screen can be as large as 1 MB. Several elements determine the file size; in addition to the length of the video, these include frame rate, image size, and color depth.

Frame rate Earlier you learned that animation is an illusion caused by the rapid display of still images (frames). Television and movies play at 30 fps, but acceptable playback can be achieved with 15 fps.

Image size A standard full-screen resolution is 640 × 480 pixels, but often video is more appropriately displayed in a window that is one-fourth (320 × 240) the size of the full screen.

Color depth Digitized video is really made up of a series of still graphic bitmaps. Hence the quality of video is dependent on the color quality (related to the number of colors) for each bitmap. As you learned earlier, an 8-bit color depth provides 256 colors, 16-bit provides more than 64,000 colors, and 24-bit provides over 16 million colors.

Using the following formula, you can estimate the file size of 1 second of digitized video:

$$\text{fps} \times \text{image size} \times \text{color depth} \div 8 = \text{file size}$$

Thus 1 second of video at a frame rate of 15 fps, with an image size of 320 × 240 and a color depth of 24 bits, would equal a file size of 3.5 MB. This means that a CD could hold only three minutes of digitized video with the stated frame rate, image size, and color depth.

Although it might be desirable to run several minutes of photorealistic full-screen video at 30 fps, it may not be feasible. The file size would be prohibitive, and the current playback multimedia systems would not support the processing power required. Thus the use of video becomes a trade-off between quality and file size.

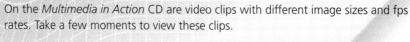

interactive exercise

On the *Multimedia in Action* CD are video clips with different image sizes and fps rates. Take a few moments to view these clips.

1. Start the CD.
2. Choose Video from the contents screen and read the instructions.
3. Click on the questions button and review the questions.
4. Click on the demos button and choose Video Clips.
5. After viewing the demo, respond to the questions previously reviewed.

What constitutes acceptable quality depends on the purpose of the title and the intended audience. Reducing the color depth below 256 colors yields a markedly poorer-quality image. Also, reducing the frame rate to below 15 fps causes a noticeable and distracting jerkiness that is usually unacceptable. Changing the image size and compressing the file therefore become primary ways of reducing file size.

You usually do not need to show full-screen video in a multimedia title, because it is only one of several elements that appear on-screen. Other elements might include text and navigational buttons. The video can therefore be played in a window as small as one-quarter or even one-sixteenth of the screen. One technique is to use video for the parts of an object that are changing. For example, assume you want to show a dinner table with a lit candle. The flame of the candle is the only moving part of the screen. Instead of creating a video of the entire table with the candle, you could use the table as a background still image and videotape only the flame. Then you could play the flame in a window at the tip of the candle, giving the impression that the entire screen is one video. The video could be played continuously to create the motion of a flickering candle. In most cases, a quarter-screen image size (320 × 240), an 8-bit color depth (256 colors), and a frame rate of 15 fps is acceptable for a multimedia title. And even this minimum results in a very large file size.

VIDEO COMPRESSION

Because of the large sizes associated with video files, video compression/decompression programs, known as **codecs**, have been developed. These programs can substantially reduce the size of video files, which means that more video can fit on a single CD and that the speed of transferring video from a CD to the computer can be increased. There are two types of compression: lossless and lossy. **Lossless compression** preserves the exact image throughout the compression and decompression process. An example of when this is important is in the use of text images. Text needs to appear exactly the same before and after file compression. One technique for text compression is to identify repeating words and assign them a code. For example, if the word *multimedia* appears several times in a text file, it would be assigned a code that takes up less space than the actual word. During decompression, the code would be changed back to the word *multimedia*. **Lossy compression** actually eliminates some of the data in the image and therefore provides greater compression ratios than lossless compression. The greater the compression ratio, however, the poorer the decompressed image. Thus, the trade-off is file size versus image quality. Lossy compression is applied to video because some drop in the quality is not noticeable in moving images.

Certain standards have been established for compression programs, including **JPEG** (Joint Photographic Experts Groups) and **MPEG** (Motion Picture Experts Group). Both of these programs reduce the file size of graphic images by eliminating redundant information. Figure 4.13 shows how the JPEG process works. Often areas of an image (especially backgrounds) contain similar information. JPEG compression identifies these areas and stores them as blocks of pixels instead of pixel by pixel, thus reducing the amount of information needed to store the image. Compression rations of 20:1 can be

Figure 4.13

JPEG compression

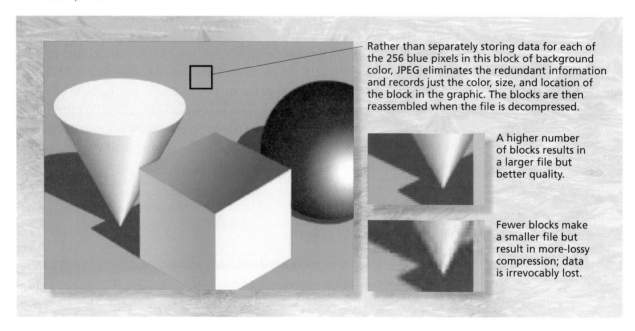

Figure 4.14

MPEG compression

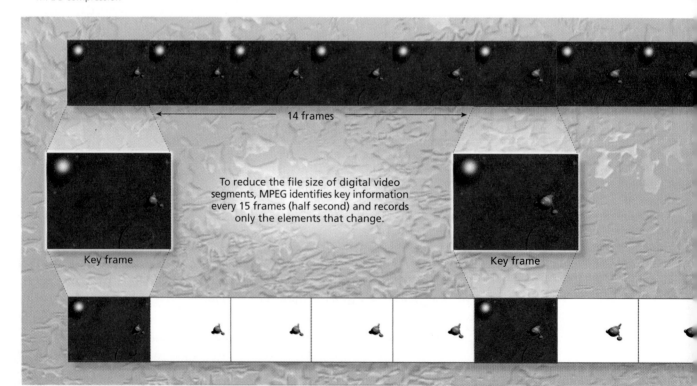

achieved without substantially affecting image quality. A 20:1 compression ratio would reduce a 1 MB file to only 50 KB.

MPEG adds another process to the still-image compression when working with video. MPEG looks for the changes in the image from frame to frame. Figure 4.14 shows, in a simplified way, how this works. Key frames are identified every few frames, and the changes that occur from key frame to key frame are recorded.

MPEG can provide greater compression ratios than JPEG, but it requires hardware (a card inserted in the computer) that is not needed for JPEG compression. This limits the use of MPEG compression for multimedia titles, because MPEG cards are not standard on the typical multimedia playback system.

Two widely used video compression software programs are Apple's QuickTime (and QuickTime for Windows) and Microsoft's Video for Windows. QuickTime is popular because it runs on both Apple and Windows-based computers. It uses lossy compression coding and can achieve ratios of 5:1 to 25:1. Video for Windows uses a format called Audio Video Interleave (AVI) which, like QuickTime, synchronizes the sound and motion of a video file.

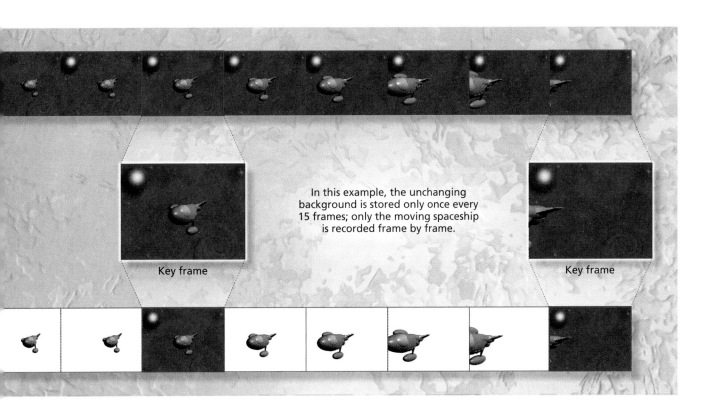

SOFTWARE FOR CAPTURING AND EDITING VIDEO

Several steps are needed to prepare video to be incorporated into a multimedia title. These include capturing and digitizing the video from some video source, such as a video camera, VCR, TV, or videodisc; editing the digitized video; and compressing the video. Some software programs specialize in one or the other of these steps, and other programs, such as Adobe Premiere, can perform all of them (see figure 4.15). Although capturing and compressing are necessary, it is editing that receives the most attention. Editing digitized video is similar to editing analog video, except that it is easier. For one thing, it is much quicker to access frames in digital form than in analog. For example, with analog video, a lot of time is spent fast-forwarding and rewinding the videotape to locate the desired frames; whereas with digital

Figure 4.15

Editing a video clip

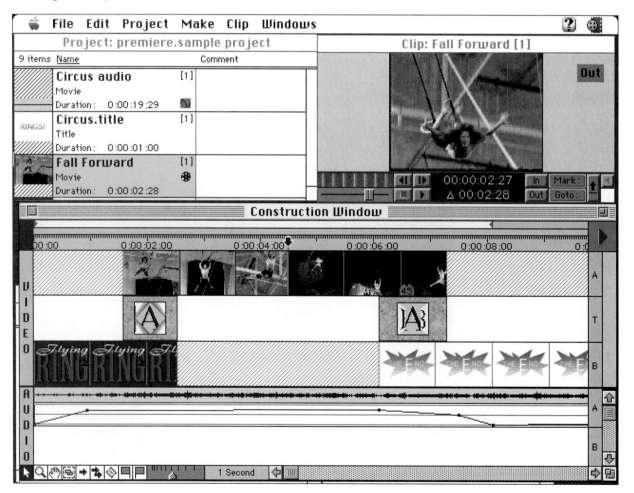

editing you can quickly jump from the first frame to the last—or anywhere in between. Removing frames or moving them to another location is as easy as the cut-and-paste process in a word processing program. The following are some other features that may be included in editing software programs:

- Incorporating transitions such as dissolves, wipes, and spins
- Superimposing titles and animating them, such as a fly-in logo
- Applying special effects to various images, such as twisting, zooming, rotating, and distorting
- Synchronizing sound with the video
- Applying filters that control color balance, brightness and contrast, blurring, distortions, and morphing

To extend what you've learned, log on to the Internet at
http://www.thomson.com/wadsworth/shuman
You will find a wide variety of resources and activities related to this chapter.

key terms

analog wave pattern
animation
cel animation
codec
frames per second (fps)
frequency
hertz (Hz)
JPEG
lossless compression
lossy compression
MIDI

modeling
morphing
MPEG
path animation
rendering
sampling
sound card
tweening
virtual reality (VR)
volume
warping

review questions

1. The three most common sample rates used to digitize sound are:

   a. _____

   b. _____

   c. _____

2. The two types of 2-D animation are:

   a. _____

   b. _____

3. The three steps involved in creating 3-D animations are:

   a. _____

   b. _____

   c. _____

4. **T F** Modeling is the final step in creating a 3-D animation.

5. **T F** A disadvantage of digital video is that it cannot be easily edited.

6. The file size of digitized video is determined by:

   a. _____

   b. _____

   c. _____

   d. _____

7. **T F** Television and movies play at 30 fps, but for most multimedia applications acceptable playback can be achieved with 15 fps.

8. **T F** JPEG and MPEG are file compression program standards.

9. What is the file size (in bytes) of 30 seconds of digitized audio in stereo using a 16-bit sample at 44 kHz?

10. What is the sample size and bit rate necessary to produce CD-quality digitized audio?

    Sample size _____ Bit rate _____

projects

1. Choose a multimedia title that does not include sound and study the title. Prepare a report suggesting how the title can be improved by using sound. Specify where in the title sound should be used, what type of sound (music, narration, sound effects, and so on), and how this will improve the title.

2. Choose a multimedia title that does not include video and study the title. Prepare a report suggesting how the title can be improved by using video. Specify where in the title video should be used and how this will improve the title.

3. Choose a multimedia title that does not include animation and study the title. Prepare a report suggesting how the title can be improved by using animation. Specify where in the title animation should be used and how this will improve the title.

4. Choose a multimedia title that includes two of the following elements: sound, animation, and video. Study the title and prepare a report indicating why these elements are important to the title. Include in your report answers to the following:

   - How do these elements enhance the title?
   - What would be the effect of eliminating some or all of them?
   - What would be the effect of substituting other elements (such as graphics for video or text for sound)?

Multimedia Authoring Programs

AFTER COMPLETING THIS CHAPTER YOU WILL BE ABLE TO:

- Describe the differences between presentation and stand-alone titles
- Compare the various categories of authoring software
- Discuss how scripting is used

A**UTHORING programs** are used to create multimedia titles. They help the developer do all of the following tasks:

- Produce content with paint, text, and animation tools
- Design screen layout using templates
- Create interactivity
- Incorporate text, sound, video, animation, and graphics
- Create hyperlinks

Multimedia authoring programs vary significantly in the features they provide and in their cost and ease-of-use. There are dozens of authoring programs to choose from, and various ways to compare them. These include the following criteria:

- Platform (Mac and/or PC) used for development and playback
- Way the developer works with them and ease-of-use
- Features (paint tool, animation tool, programming language)
- Price
- Learning curve
- Ability to develop multimedia applications that can be delivered via the Internet

Two ways in which multimedia is used are in presentations and as stand-alone titles. It is helpful to distinguish between these uses, because the type of application affects the design of the title, the cost, the development time, and often the authoring program used.

Multimedia Presentations

Multimedia presentations involve a presenter and an audience of one or more persons. Examples include the following:

- A college professor lecturing on the art collection of the Hermitage Museum in St. Petersburg and using a computer to display various paintings as well as biographies of the artists
- A sales presentation in which a representative uses a computer to display the company's new product line, including animations of how the products work

- A corporate CEO making a presentation at the annual stockholders' meeting and using a computer to display highlights of the past year, pictures of corporate officers, and financial data

In these cases, the presenter has control over the multimedia title, and it is primarily a one-way, linear communication process. If the multimedia title allows the presenter to quickly navigate through the contents of the title, a great deal of interactivity can be built into the presentation. For example, a sales representative may be making a pitch to a prospective client that starts with a relatively linear presentation about her company's history and product line. Then she may ask a few questions and, based on the responses, jump to information about pricing or product availability or to video clips of testimonials from current customers. These types of presentations are useful when the presenter wants to utilize the power of multimedia while maintaining control of the presentation. Many of these presentations are similar to a slide show and can be developed easily, quickly, and inexpensively.

Stand-alone Applications

Stand-alone titles are those that are intended for use by individuals in a one-on-one situation. Examples include the following:

- A computer-based simulation of a biology lab procedure in which students learn how to dissect a frog
- An informational kiosk located in a shopping mall, with which customers can locate various stores and view product lines
- An interactive CD-based encyclopedia located in a library
- A CD-based sales catalog distributed through the mail to potential customers
- A corporate training CD used to teach employees how to deal with angry customers
- A solve-the-mystery adventure game distributed on CD

The primary differences between presentation and stand-alone titles are who has control and the amount of interactivity that is involved. A major advantage of stand-alone titles is that the user has control and can determine what to view and review based on his or her needs. It is possible, however, for developers to build into these titles conditions that the user must satisfy before proceeding along a certain path. For example, in an adventure game the user might be required to solve a puzzle before being allowed through a certain door; or in an educational title the user might be required to take a pretest that determines which tutorial can be accessed.

All stand-alone authoring programs have one thing in common: the ability to create hyperlinks. **Hyperlinking** is the process of establishing a relationship (link) between two elements or objects within a title. An example would be a word or phrase, called a hotword or **hypertext**, that, when clicked, causes the program to jump to another screen. Because control is turned over to the user, several design, navigation, and "what-if" issues must be addressed. This makes these titles generally more difficult, expensive, and time-consuming to develop than presentation-type titles.

How Authoring Systems Work

Multimedia authoring systems can be categorized by the way in which they work—that is, the **metaphor** used. These include the electronic slide show, the card stack or book, icon-based programs, and time-based programs using a movie metaphor.

ELECTRONIC SLIDE SHOWS

Giving an **electronic slide show** presentation is similar to using overhead transparencies or traditional photographic slides. Although any authoring program can create an electronic slide show, some programs are specifically designed for developing them. Programs such as Microsoft PowerPoint, Adobe Persuasion, Asymetrix Compel, and Software Publishing's Harvard Graphics use a slide show metaphor. Figure 5.1 shows the PowerPoint screen used to specify a sound to be played. Figure 5.2 shows the thumbnail view of several slides in a PowerPoint presentation.

These types of programs have several advantages. They are relatively inexpensive, easy to learn, and easy to use. Developers of multimedia titles are generally familiar with slide shows, and it is easy to visualize each screen being a separate slide. These programs provide templates with different background colors and graphics and allow you to incorporate all of the multimedia elements, including video, animation, and sound. Some programs have hyperlinking capabilities that allow the user to navigate to any part of the application. Most of the programs run on both the Macintosh and Windows platforms, and many have runtime modules which allow the presentation to be played on computers that do not have the program. The primary disadvantage of slide show presentations is their linear, noninteractive nature.

Chapter 5 Multimedia Authoring Programs

Figure 5.1

A PowerPoint screen used to specify a sound to be played

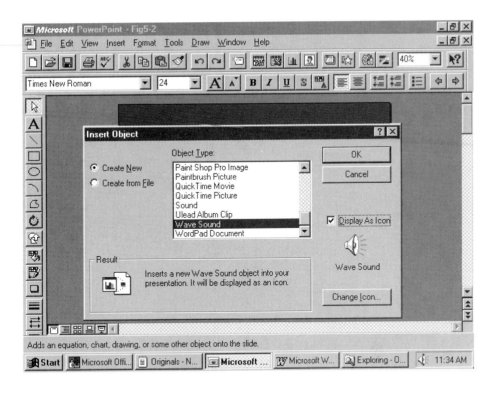

Figure 5.2

The thumbnail view of several slides

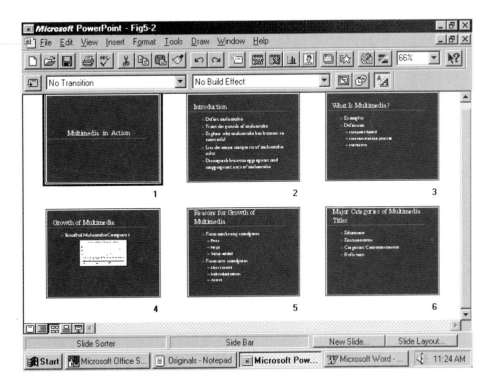

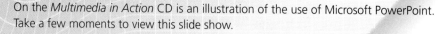

interactive exercise

On the *Multimedia in Action* CD is an illustration of the use of Microsoft PowerPoint. Take a few moments to view this slide show.

1. Start the CD.
2. Choose Authoring from the contents screen and read the instructions.
3. Click on the questions button and review the questions.
4. Click on the demos button and choose PowerPoint.
5. After viewing the demo, respond to the questions previously reviewed.

THE CARD STACK AND BOOK METAPHORS

One of the first multimedia authoring programs was HyperCard, which ran on Apple computers. HyperCard uses a **card stack metaphor**. Cards are developed that have different elements associated with them and are put in stacks as shown in figure 5.3. You can link the cards by allowing the user to click on buttons or other elements and jump to a different card in the stack. A similar type of program uses a **book metaphor**, in which each page represents a different screen, and the pages combine to make up a book. ToolBook by Asymetrix is an example of this type of program (see figure 5.4). There are two levels in this program: author and reader. The author level allows you to create the title, whereas the reader level allows you to interact with it as a user. ToolBook provides a paint program for basic drawing, as well as a path animation feature for creating 2-D animations.

The advantages of using the card- and book-type authoring programs include the ease in understanding the metaphor, and the straightforward relationship between what is displayed on any particular screen and what is created on a card or page (see figure 5.5). These programs are also relatively easy to use and often provide templates that can significantly shorten development time. A major disadvantage is that some of the programs run on only one platform (Macintosh or Windows). In addition, some of their features, such as animation and paint tools, are not as powerful as those in other programs.

Chapter 5 Multimedia Authoring Programs

Figure 5.3

A HyperCard stack

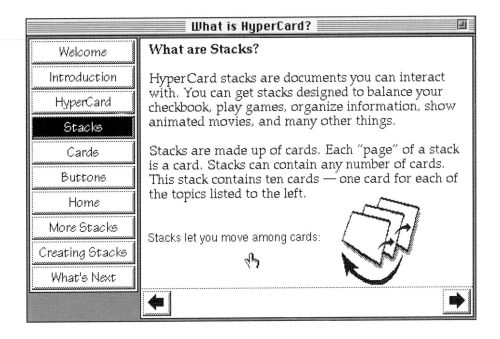

Figure 5.4

A ToolBook application

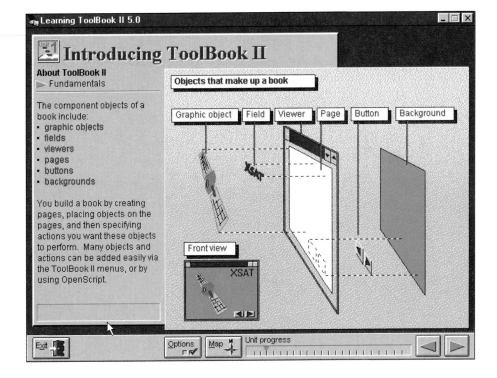

Figure 5.5

A page design from the authoring program and the screen it creates

interactive exercise

On the *Multimedia in Action* CD are two multimedia titles. One illustrates HyperCard; the other illustrates ToolBook. Take a few moments to view these titles.

1. Start the CD.
2. Choose Authoring from the contents screen and read the instructions.
3. Click on the questions button and review the questions.
4. Click on the demos button and choose HyperCard and ToolBook.
5. After viewing the demos, respond to the questions previously reviewed.

ICON-BASED AUTHORING PROGRAMS

With ***icon-based programs***, you use symbols in a flowchart scheme as shown in figure 5.6. Each icon represents a particular event. For example, the Wait icon stops the process until the user clicks the mouse button or presses a key, or a specific amount of time passes.

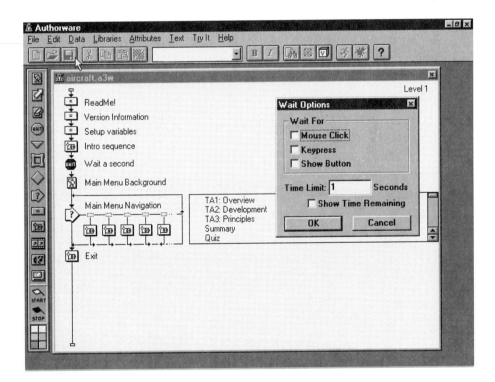

Figure 5.6

An icon-based authoring program

An advantage of icon-based programs is that you can easily see how a title is structured, that is, the flow of a program and especially the branching. This makes it easy to edit and update the program by merely adding or deleting icons representing content or events. There are, of course, some disadvantages to authoring with icon-based programs. They are less intuitive than the other programs discussed in this chapter, and they can be fairly difficult to learn. They are often also more expensive.

interactive exercise

On the *Multimedia in Action* CD is a multimedia title developed using Authorware, an icon-based authoring program. Authorware runs on both Macintosh and Windows-based computers. Take a few moments to view this title.

1. Start the CD.
2. Choose Authoring from the contents screen and read the instructions.
3. Click on the questions button and review the questions.
4. Click on the demos button and choose Authorware.
5. After viewing the demo, respond to the questions previously reviewed.

TIME-BASED AUTHORING PROGRAMS

Time-based programs use a movie metaphor. That is, like a movie on videotape, you start the multimedia title and it plays until some action causes it to pause or stop. Although this may seem contrary to the nonlinear nature of multimedia, these programs allow for branching to different parts of the "movie," and any amount of user control and interactivity may be built in. Figure 5.7 shows a popular time-based program, Macromedia Director. Instead of cards or book pages, Director movies are made of a series of individual frames. Each frame consists of up to 48 objects such as graphics, buttons, and text placed on a stage (the computer screen). As the movie is played, frames are displayed, revealing their elements.

Because time-based authoring programs work by displaying a series of frames, they are especially good for creating animations. The more sophisticated products have powerful programming features and are dual platform; that is, you could develop a multimedia title using one type of

Figure 5.7

A time-based program

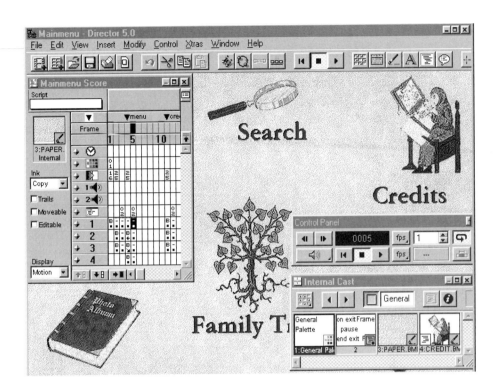

computer (Macintosh) and play it back on another type (Windows). These programs tend to be more costly and usually have a steep learning curve for the advanced features.

interactive exercise

On the *Multimedia in Action* CD is a multimedia title developed using Macromedia Director, a time-based authoring program. Director runs on both Macintosh and Windows-based computers. Take a few moments to view this title.

1. Start the CD.
2. Choose Authoring from the contents screen and read the instructions.
3. Click on the questions button and review the questions.
4. Click on the demos button and choose Director.
5. After viewing the demo, respond to the questions previously reviewed.

Programming Languages

Although many authoring tools have been developed specifically for creating multimedia titles, traditional ***programming languages*** are also used. Among the most common are C++ and Visual Basic. C++ is perhaps the most popular programming language. It is used by software companies to create word processing and spreadsheet programs, and by other businesses to develop specific applications, such as inventory control and payroll programs. The advantage of using a programming language is its flexibility. Because you are writing the code (the instructions carried out by the computer), you can be very specific in tailoring the application. This is especially important when you are developing titles that may run on different operating systems, such as those in foreign countries like Japan. Because authoring programs are designed specifically for creating multimedia applications, it is easier and less time-consuming to develop titles with them than with traditional programming languages.

SCRIPTING

All multimedia titles are based on programming code. The question is: How is the code generated? Will the authoring program generate the code automatically as the title is created, or will the developer need to write the code? Authoring programs are specifically for developing multimedia titles, so they try to make it easy for developers to produce and edit content, create hyperlinks and interactivity, and incorporate text, sound, graphics, animation, and video. Some authoring programs are promoted on the basis of their ease-of-use and the fact that you do not have to be a programmer to use them. As a developer works with these authoring programs, code is automatically generated and becomes transparent to the developer. Other authoring programs, however, come with their own programming language, often called ***scripts***. These languages allow the developer flexibility in designing the title and may be necessary for such things as accessing external media such as videodiscs; creating interaction (for example, specifying what happens when a user types a certain response to a question); or controlling the speed of an animation.

Following are examples of programming code from two different authoring programs. These are simple examples that illustrate how easy it is to understand some code.

ASYMETRIX TOOLBOOK'S OPENSCRIPT

Example 1

ToolBook uses a book metaphor wherein each page of the book is a screen. Figure 5.8 shows two screens in a ToolBook application. In the first screen, the user selects an option by clicking on a button. Each button is scripted to cause an action (in this case, branch to a different page). If the user clicks on the button labeled True, the second screen in figure 5.8 is displayed. The following is the script that causes this action:

```
to handle buttonClick
    go to page "Correct"
end buttonClick
```

There are three lines of script. The first line indicates that the script is executed when the mouse button is clicked (assuming the mouse pointer is on the button named True). The second line causes the program to go to a page named Correct). The third line ends the script.

The ToolBook program provides two ways to create a script. You can type it or you can use a dialog box to specify a link (see figure 5.9), which causes the program to write the script.

Figure 5.8

Example of hyperlinking two screens with a button

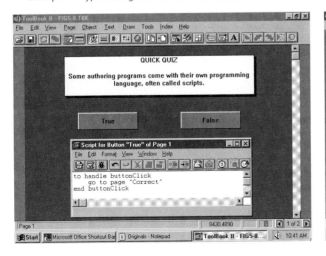
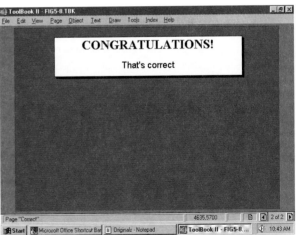

Figure 5.9

Using a dialog box to generate the script

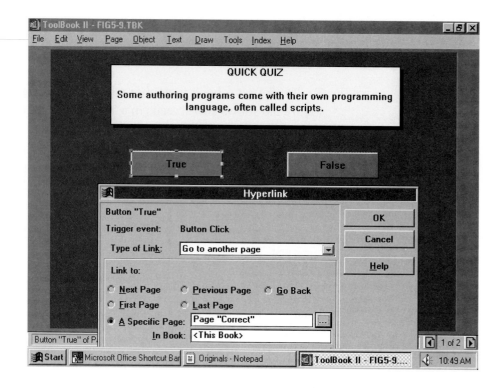

Example 2

Figure 5.10 shows an animation that can be created by quickly displaying successive pages. Toolbook has a Flip command that flips through the number of pages you specify. The pages are displayed so quickly (fractions of a second) that an animation is created. In this case, the user clicks on the word *Enter* on the first page, and the next four pages are displayed one at a time. The following is the script for this action:

```
to handle buttonClick
    flip 4 pages
end buttonClick
```

Figure 5.10

An animation created using a script

MACROMEDIA DIRECTOR'S LINGO

Example 1

As mentioned earlier, Director uses a movie metaphor. When you start a movie, a playback head moves from left to right one frame at a time, displaying whatever is in the frame. The playback head can be stopped at any frame, and the user is allowed to choose how to proceed. Based on the user's actions, the program can branch to another frame. This is similar to jumping to another page in ToolBook. Figure 5.11 shows the frames of a movie and the playback head stopped at frame 5. Figure 5.11 also shows what the user sees at this point. The following script causes the playback head to jump to another frame (20) when the user clicks on the graphic called Gallery (see figure 5.12):

```
on mouseUp
    go to frame 20
end mouseUp
```

Example 2

In this example a function called rollOver is used. When the mouse pointer rolls over (points to) an object, some action occurs. In this case, when the mouse pointer rolls over the car, the words *Grand Cherokee* are displayed as

Figure 5.11

The frames of a movie with the playback head stopped at frame 5

Figure 5.12

Frame 20 displays this screen

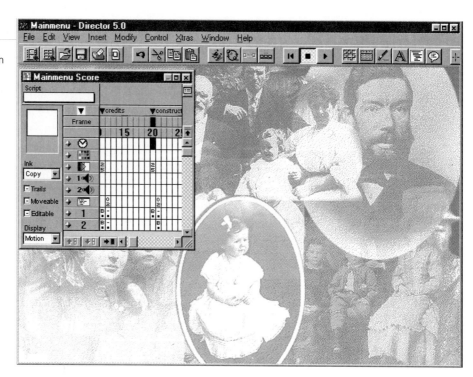

shown in figure 5.13. After the rollOver code is the number of the object, 5. The first line of code says that if the mouse pointer is over object number 5, continue to the next line of code. The second line puts the words *Grand Cherokee* into a field called Message, which has been previously designated to be an area on-screen that will display text. The last line of code ends the if statement.

```
if rollOver(5) then
    put "Grand Cherokee" into field "Message"
end if
```

CUSTOM CODES AND THIRD-PARTY DEVELOPERS

Programming can be a costly and time-consuming process requiring a high level of expertise. Companies such as Macromedia, which sells Director, and third-party developers provide modules consisting of code that can be "plugged-in" to a Lingo script. One set of modules, called XObjects, allows Director movies to interact with external software and hardware such as CD-ROM players. Another set of modules, called Xtras, extends Director's functionality by allowing developers to create custom in-house code that works with 3-D and database programs. One source of custom codes is user

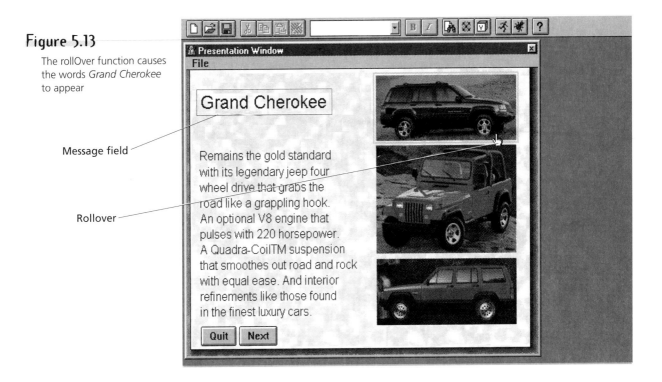

Figure 5.13

The rollOver function causes the words *Grand Cherokee* to appear

Message field

Rollover

groups, often found on the Internet, composed of individuals who are willing to share their custom codes.

WEB-BASED AUTHORING TOOLS

As the World Wide Web has become more viable as a distribution alternative for multimedia, companies are adapting their authoring tools to accommodate the Web. Asymetrix and Quest provide versions of their authoring programs specifically designed for developing multimedia titles to be delivered over the Internet. Macromedia provides a utility program, called Shockwave, that allows an Internet user to play Director movies; and Microsoft provides its ActiveX controls that allow movies, animations, and sound to be delivered over the Internet.

The decision of which authoring system to use is based on the target playback system, the desired features, the development timeline, the budget, and the developer's expertise.

To extend what you've learned, log on to the Internet at

http://www.thomson.com/wadsworth/shuman

You will find a wide variety of resources and activities related to this chapter.

key terms

authoring programs
book metaphor
card stack metaphor
electronic slide show
hyperlinking
hypertext
icon-based programs
metaphor
multimedia presentation
programming language
script
stand-alone titles
time-based programs

review questions

1. List three ways authoring programs can be compared:
   a. _____
   b. _____
   c. _____

2. List two ways in which multimedia is used:
   a. _____
   b. _____

3. **T F** The primary differences between presentation and stand-alone titles are who has control and the amount of interactivity that is involved.

4. **T F** An advantage of electronic slide show programs is that they are relatively easy to learn.

5. **T F** An advantage of icon-based authoring programs is that they show the flow of a program, especially the branching.

6. **T F** Director is an example of an icon-based authoring program.

7. Explain each line of the following script:

   ```
   on mouseUp
       put "Correct" into field "Answer"
   end mouseUp
   ```

8. **T F** The rollOver command in ToolBook is used to create animations.

9. **T F** Stand-alone titles are intended for use by individuals in a one-on-one situation.

10. **T F** Time-based authoring programs use a card stack metaphor.

projects

1. Select two authoring programs (at least one that is used to create stand-alone titles) and prepare a report that evaluates them on the following criteria:
   - Platform(s) used for development and playback
   - Features
   - Price
   - Difficulty of learning/ease-of-use
   - Metaphor used

   List a few lines of script used in one of the programs (other than the examples used in this chapter) and explain what is happening.

   Prepare an oral presentation of your report and be ready to present it to your class.

2. Visit a retail store that sells CD-ROM titles and study the package to determine the authoring programs and other software tools used in developing the titles. Determine if there is a correlation between the type of title (game, educational, reference, and so forth) and the authoring program used. Develop a report of your findings.

   Prepare an oral presentation of your report and be ready to present it to your class.

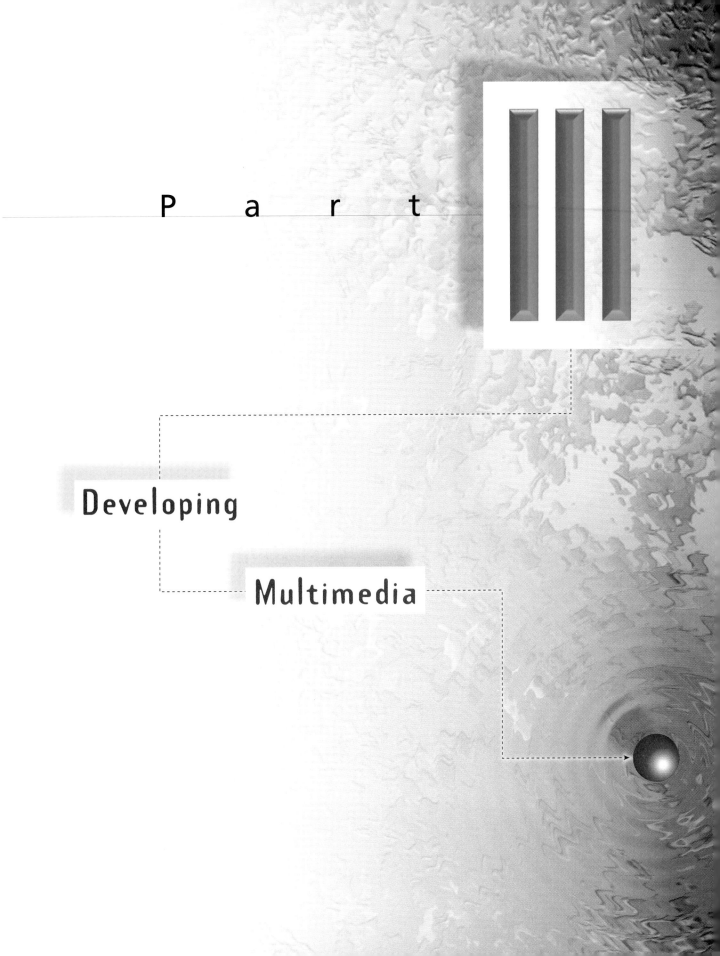

Part III

Developing Multimedia

Developing Multimedia Titles

AFTER COMPLETING THIS CHAPTER YOU WILL BE ABLE TO:

- List and describe the steps in developing a multimedia title

- Give an example of how the intended audience affects the design and content of the title

- Describe the relationship between the storyboard and the navigation scheme

IN this chapter you will learn the process that developers use to create multimedia titles. Like many other processes that require the use of the computer (programming, report writing, computer-aided design), the tendency is to jump right into the project by loading a graphics or authoring program and beginning to create the title. This tendency is reinforced because computers and computer software have become so easy to use and are so forgiving. If you make a mistake while typing a report using a word processing program, you can easily correct it. There are undo and redo features as well as spelling checkers and debugging routines. Also, you feel that you are accomplishing something because it is "on the computer." In reality, however, developing multimedia can be extremely complex, costly, and time-consuming. A single commercial-quality CD title could take months to develop; involve the coordination of people, resources, and budgets; and cost hundreds of thousands of dollars. The "wing it" approach invariably results in less effective and more costly titles. The rule-of-thumb for multimedia development is 80 percent planning and 20 percent production. Of the following nine steps, only two focus on production. Pursuing a well-defined development process that places the actual production in perspective will help ensure a successful project.

Steps in Developing Interactive Multimedia

The steps in multimedia development can be grouped in several ways, such as preproduction, production, and postproduction; or design, production, and distribution. The phases presented here are planning, creating, and testing. Multimedia distribution is covered in chapter 10. The following steps are presented as a numbered series. In practice, however, there is a great deal of overlap and rework among them. For example, content development may overlap with authoring; and testing, although it is listed as the last step, is an ongoing process throughout the project.

The three phases of multimedia development and the steps they involve are outlined below.

Phase 1—Planning

- Step 1: Developing the concept
- Step 2: Stating the purpose
- Step 3: Identifying the target audience
- Step 4: Determining the treatment
- Step 5: Developing the specifications
- Step 6: Storyboard and navigation

Phase 2—Creating

- Step 7: Developing the content
- Step 8: Authoring the title

Phase 3—Testing

- Step 9: Testing the title

The Planning Phase

STEP 1: DEVELOPING THE CONCEPT

"What, in general, do we want to do?"

Every multimedia project originates as an idea. The idea for what was to become a series of very popular children's CD titles came from a story that a mother created for her own children. The process for generating ideas can be as unstructured as brainstorming sessions or as formal as checklists with evaluation criteria. Companies that rely on providing a continuous stream of products might have a process for generating new product ideas that includes asking questions about the current product line. Such a series of questions might consist of the following:

- How can we improve it (make it faster, use better-quality graphics or updated content)?

- How can we change the content to appeal to a different market (consumer, education, corporate)?

- How can we take advantage of new technologies (virtual reality, speech recognition)?

- How can we make it disposable?
- How can we change it from a single title to a series?
- How can we repackage or repurpose our content (books, movies, games, reference materials, brochures, magazines)?

One of the tenets of marketing is to "find a need and fill it." A father's and mother's need to be thought of as responsible parents helped spawn an entire industry of children's educational titles. The need of educators to address different learning styles and motivate students by giving them more control over the learning process has resulted in hundreds of interactive tutorials, simulations, and testing titles. Publishers' and entertainment companies' need to increase profits has resulted in CD versions of books and movies.

Because manufacturing is market driven, companies rely on customer and employee feedback to help generate new ideas. This could be in the form of product support lines (help desks) or feedback from retailers on why products are returned.

The following are examples of typical questions and comments for generating ideas for multimedia titles.

Professor: "My students have a difficult time visualizing the theory of DNA replication. If they could experience the process through a computer simulation, it might help them understand."

Marketing executive: "I hear a lot about the Internet, the World Wide Web, and home pages; could we be using these to promote our products?"

Personnel director: "Each new employee we hire has to be tested for computer skills, specifically word processing. Then, if they need training, I must either set up an expensive one-on-one training session, or reimburse them for a outside course. If this process could be computerized, we could reduce our training costs considerably."

Publishing or movie executive: "We own all this content (film archives, manuscripts, screenplays, documentaries). What other ways can we use it?"

Ideas can provide the vision, but they must be presented in a way that can guide the development process. That is, they must be stated as clear, measurable, and obtainable objectives.

STEP 2: STATING THE PURPOSE

"What, specifically, do we want to accomplish?"

Once a concept has been developed, project goals and objectives need to be specified. This is perhaps the most critical step in multimedia planning. **Goals** are broad statements of what the project will accomplish, whereas **objectives** are more precise statements. Goals and objectives help direct the development process and provide a way to evaluate the title both during and after its development. Because multimedia development is a team process, objectives are necessary to keep the team focused, on-track, on budget, and on time. They need to be stated in measurable terms, and they need to provide for a timeline. The following are examples of goals and objectives.

GOALS
Broad statements of what a project will accomplish:

"Be the leader in educational CDs"

"Create products that take advantage of emerging technologies (such as the Internet)"

"Use multimedia to reduce our training costs"

"Create a line of award-winning multimedia biographical titles by repurposing existing content (film biographies of historical figures)"

"Produce the best CD title on fly fishing"

The goal for a specific multimedia title must fit within the overall mission of the company or organization. For example, if the company's mission is to create and distribute children's educational titles, establishing a goal of developing the "best" children's encyclopedia would be consistent with that mission. However, *best* is a subjective term that provides little guidance in developing a title. The following are objectives that would be useful to a development team.

OBJECTIVES
Precise statements of what a project will accomplish:

"To develop an entertainment title based on the book *Tracks,* which chronicles one woman's journey across the continent of Australia. The title will include an interactive map that shows her progress and allows the user to view photographs and text about any selected map location. Sound clips will provide narration of the author's adventures in her own voice. The title will be rich with photographs of the outback and music native to Australia. The product will place in the top five for its category at the annual CD awards this year."

"To develop a corporate image title that traces the restoration of the landmark Paramount theater. The title will allow the user to select from various categories such as History of the Theater, Premiere Plays, Architectural Drawings, Financial Support, and Opening Night Gala. The title will feature narration by the chairman of the Board of Trustees and will include a virtual reality tour that allows the user to navigate through the theater. The title will be completed for distribution on opening night."

STEP 3: IDENTIFYING THE TARGET AUDIENCE

"Who will use the title?"

The more information a developer has about potential users, the more likely a title can be created that will satisfy the users' needs and be successful. A child's way of interacting with a computer and computer programs is different than an adult's. Children are more inclined than adults to respond to the elements of exploration and surprise, and to have a guide (character that they view as a helper or friend) lead them through a title.

Audiences can be described in many ways, in terms of **demographics** (location, age, sex, marital status, education, income, and so on) as well as lifestyle and attitudes. Developers must determine what information is needed and how specifically to define the audience. There is a trade-off between the size of an audience and a precise definition of it. Companies want to identify as large an audience as possible in order to maximize potential sales. The larger the audience, however, the more diverse its needs and the more difficult it is to "give them what they want." For example, a company might want to develop a title on earthquakes, the objective being to teach people how to prepare for an earthquake and what to do if caught in one. At first it may seem that this would be useful for anyone, but a person living in the Midwest might not see the need to buy the product. In this case, the target audience is defined, to some degree, by geographical location. Alternately, the company could consider creating a series of disaster preparedness titles (earthquakes, hurricanes, tornadoes, floods) that would have the same basic structure but the content would vary. This would increase the potential market and keep down the cost of developing each title.

STEP 4: DETERMINING THE TREATMENT

"What is the 'look and feel'?"

Taken together, the concept, objectives, and especially the audience will help determine how the title will be presented to the user. *"**Look and feel**"* can include such things as the title's tone, approach, metaphor, and emphasis.

Tone Will the title be humorous, serious, light, formal? Many multimedia titles intended for home use, such as games and recreational titles, include humor, whereas those intended for business use are more serious in their tone. Titles intended for children tend to be whimsical, whereas training titles are generally straightforward and conservative. Titles that include original graphics rather than clip art, 3-D rather than 2-D animation, and video clips with sound rather than still images might help a company project a progressive, high-tech, well-funded corporate image.

Approach How much direction will be provided to the user? Some titles, especially children's games and interactive books, focus on exploration. The child is presented with a scene, like the one shown in figure 6.1, with little or no instructions. The child clicks on various objects (chimney, mailbox, sky, tree) and some action occurs. For example, clicking on the sky causes an airplane to appear. Other titles, especially adult-education applications,

Figure 6.1

A children's title with the focus on exploration

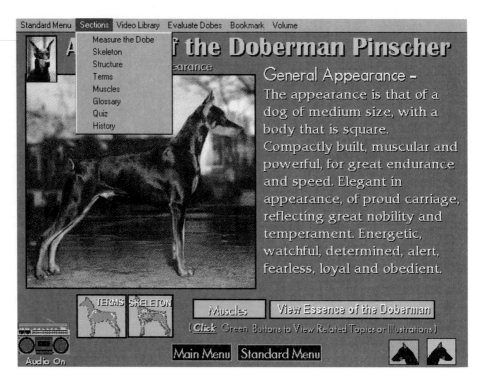

Figure 6.2
A title that provides a menu

provide a great deal of direction. Users are provided with menu choices, as shown in figure 6.2, and generally must follow a predetermined path in order to complete the title. Another aspect of approach is deciding how much help to provide and in what form. Some titles provide a "host" or "guide" that is available to assist the user. Microsoft's Explorapedia, for example, a children's encyclopedia, has a tadpole named Thaddeus Pole—nicknamed Tad—who acts as a companion and helper as the user explores the world of nature. Other titles simply have a help button that triggers a help screen. In other cases, a main character, such as Freddie Fish shown in figure 6.3, is used as a guide.

Metaphor Will a metaphor be used to provide interest or to aid in understanding the title? The Explorapedia title uses space travel as the theme for exploring different areas of content such as Nature and People. The TOEIC test program uses a mountain-climbing metaphor in which the user starts at "base camp" and, each time a set of questions is answered correctly, proceeds to a higher camp until the summit is reached. The Odyssey humanities instructional program uses a sailing-adventure metaphor in which students choose from various island destinations (labeled by topical area such as Economics, History, and Philosophy).

Emphasis How much emphasis will be placed on the various multimedia elements? It is important to consider the significance of each element based

Figure 6.3

A title that uses a main character as a guide

on the concept, objectives, and audience for the title. Budget and time constraints, however, may ultimately dictate the relative weight placed on text, sound, animation, graphics, and video. For example, a company may want to develop an informational title that shows the features of its new product line, including video clip demonstrations of how each product works. But if the budget did not allow for the expense of creating the video segments, the emphasis would be on still pictures with text descriptions that might already be available in the company's printed catalogs.

interactive exercise

On the *Multimedia in Action* CD are demonstrations of titles that have varying treatments. Take a few moments to view these.

1. Start the CD.
2. Choose Treatment from the contents screen and read the instructions.
3. Click on the questions button and review the questions.
4. Click on the demos button and follow the instructions.
5. After viewing the demo, respond to the questions previously reviewed.

STEP 5: DEVELOPING THE SPECIFICATIONS

"What precisely does the title include and how does it work?"

Project specifications to a multimedia developer are like a movie script to a Hollywood producer. The **specifications**, or "specs," list what will be included on each screen, including the arrangement of each element and the functionality of each object (for example, what happens when you click on the button labeled Next). Specifications should be as detailed as possible. The more detailed and precise the specifications, the greater the chance of creating a title that will meet the objectives of the project on time and within budget. The goal in creating the specifications is to be able to give them to the production team and, with little further instructions, have the team create the title. In practice this doesn't happen, because of the inability to know at the start of the project exactly what is needed; it's impossible to anticipate every problem or every opportunity as the project progresses. For example, a college informational kiosk project was originally specified to play back on a stand-alone computer with a videodisc. Near the end of the development process, however, the college decided to deliver the information via a home page on the World Wide Web. It was decided to remove the video segments and some of the larger graphic files that would not download quickly enough to a user's computer.

Although specs will of course vary from project to project, there are certain elements that should be included in the specifications for all titles. These are listed below and are explored in more detail in the discussion that follows.

- Target playback system(s)
- Elements to be included
- Functionality
- User interface

Target playback systems   The decision of what computers to target for playback is usually not difficult and in some cases not even discretionary. For example, an instructor who is developing a multimedia presentation would be confined to the playback system set up in the classroom; a sales representative might be restricted by the model of laptop computer that she carries; or a person developing a title that runs on a kiosk would be restricted to the kiosk hardware.

Companies developing commercial titles intended for the business market are guided by the fact that 80 percent of desktop computers are Windows-based systems. Companies developing for the K–12 education market, on the other hand, are faced with a majority of the computers being Apple systems. Household computers are more evenly divided, so many companies targeting the home market develop for both platforms. Even then there is the need to specify the models and operating systems (such as MPC2, Windows 95, Macintosh with 68040 processor, Apple System 7.x, and so on). The platform issue is especially critical when an objective is to sell the product in a foreign country.

Elements to be included The specifications should include, as much as possible, details about the various elements that are to be included in the title. If sound is used, should it be recorded at 44 MHz, 16 bit, stereo? Should the resolution for the graphics be 8 bit, 256 colors? Should video be designed to play back at 15 frames per second? At what size? What are the sizes of the various objects such as photos, buttons, text blocks, and pop-up boxes? What fonts, point sizes, and type styles are to be used? What are the colors for the various objects? The multimedia elements chosen may require other specs. For example, if a narration or voice-over is used, a script would be part of the specifications.

Functionality Objects such as text, graphics, buttons, and hypertext are often part of a multimedia title. The specifications should include how the program reacts to an action by the user, such as a mouse click. For example, clicking on a door (object) might cause the door to open (an animation); a doorbell to ring (sound); an "Exit the program?" message to appear (text); or an entirely new screen to be displayed. In addition, specifications should include how the object itself changes based on a user action. For instance, when the user clicks on a button (object), how does the button change? The user needs feedback that the button has been selected, such as the button appearing "pressed." If no feedback is given, the user might click on the button again, resulting in the undesirable effect of "jumping" to the wrong screen. Figure 6.4 shows three screens with a Next button that is used to display the screens in sequence. Because the Next buttons are positioned in the same place on every screen, it might be possible for the user to click twice on the Next button on screen 1, and the program would register one click for screen 1 and one click for screen 2. Thus the user would go from screen 1 to screen 3. Although there are some programming techniques to help avoid this, there are ways of giving feedback so the user knows an object has been selected. For buttons, this could be a clicking sound and a change in its appearance (button shown in its pressed state or in a different color). Whatever planned change occurs to the object would become part of the specifications.

Figure 6.4
Three screens with the Next button used to navigate among them

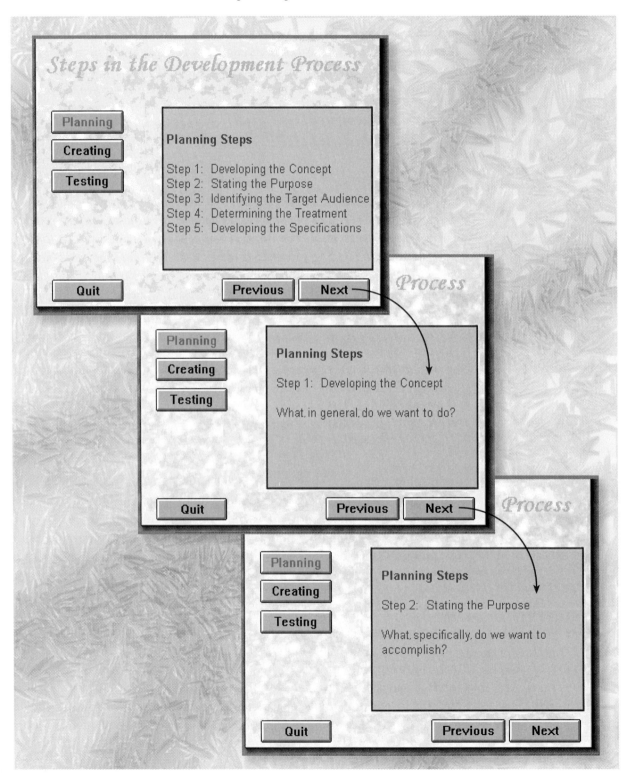

User interface The *user interface* involves designing the *appearance*—how each object is arranged on the screen—and the *interactivity*—how the user navigates through the title. The design issues relating to multimedia are extremely important, and the next chapter is devoted to studying them.

STEP 6: STORYBOARD AND NAVIGATION

"What do the screens look like and how are they linked?"

Multimedia borrows many of its development processes from movies, including the use of storyboards. A ***storyboard*** is a representation (often in the form of hand-drawn sketches) of what each screen will look like and how the screens are linked. The storyboard serves multiple purposes:

- To provide an overview of the project
- To provide a guide (road map) for the programmer
- To illustrate the links among screens
- To illustrate the functionality of the objects

Figure 6.5 shows a storyboard. Notice that it comprises hand-drawn sketches on 8½-by-11-inch paper turned sideways to more closely represent the dimensions of a computer screen. Also notice that each frame represents

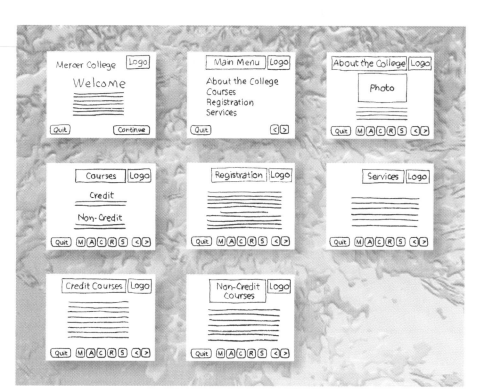

Figure 6.5
A storyboard

Figure 6.6

The linking of various screens

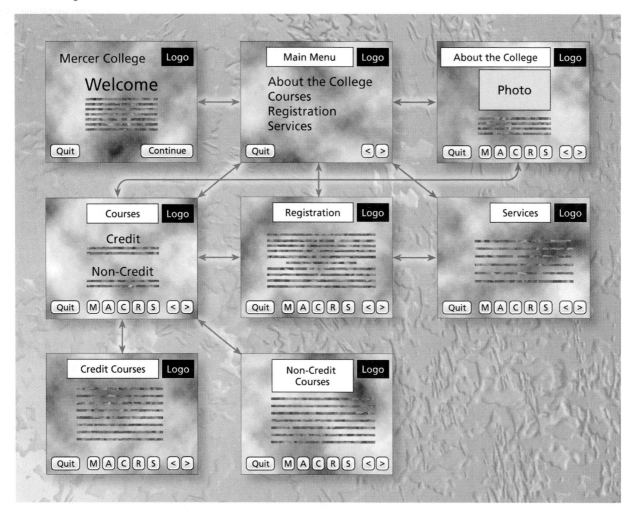

one screen of the title and presents a rough layout of the elements to be displayed on the screen and their approximate size and location. It is not necessary at this point to have decided on the exact content, such as a particular photo or graphic. It is important to show where text, graphics, photos, buttons, and other elements would be placed. Thus the storyboard could include placeholders for the various elements.

Another important feature of the storyboard is the navigation scheme. One of the most significant aspects of multimedia is the nonlinear interactivity. The linking of screens through the use of buttons, hypertext, and hot spots allows the user to jump from one screen to another. The multimedia developer decides how the various screens will be linked, and this is represented on the storyboard. In this way, problems with the navigation scheme can be determined before the programming begins. Figure 6.6 shows the primary

Figure 6.7

A sequential navigation scheme

links for the various screens. In some cases, the linking is too complex to lend itself well to a storyboard display. In such instances the programmer would rely on the specifications to indicate the navigation scheme.

Navigation schemes can be set up in a variety of ways, including sequential, topical, and exploratory. A ***sequential navigation*** scheme takes the user through a more or less controlled, linear process. Examples are games with a story line that has a beginning, a middle, and an end; books that are repurposed as multimedia titles; slide show presentations; and instructional tutorials that require the student to move through the material step-by-step (see figure 6.7). Sequential titles often have buttons (Next, Forward, Continue, Previous, Back) or graphics (arrows, pointing fingers) as navigational aids. To keep the user on track, interactivity might be limited

Figure 6.8

A topical navigation scheme

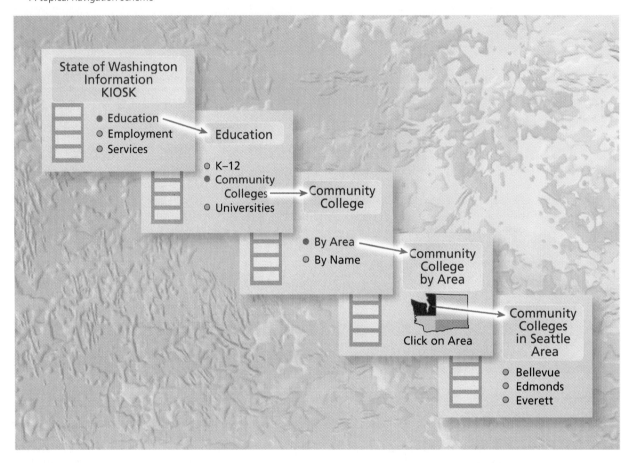

to clicking on objects that cause an action, such as playing a sound or animation or displaying a graphic or text.

A *topical navigation* scheme allows the user to select from an array of choices or even search for specific information. Examples are multimedia encyclopedias, interactive shopping catalogs, and information kiosks. Topical navigation schemes often have several layers as shown in figure 6.8. This requires special attention when designing the interface to make sure that users understand where they are and what they can do.

An *exploratory navigation* scheme provides little structure or guidance. It relies on user interaction, usually the clicking of objects displayed on the screen. Many games, directed at both children and adults, use some form of exploratory navigation.

The Creating Phase

At this point in the development process, the focus changes from planning to production, including creating the content and authoring the title.

STEP 7: DEVELOPING THE CONTENT

"Creating the pieces"

The specifications, including any scripts, indicate the content to be incorporated into the multimedia title. There are numerous content issues that need to be addressed:

- What is the level of quality for the content (photorealistic graphics, stereo sound)?

- How will the content be generated (repurpose existing content, hire content experts to write text, employ graphic artist and other professionals)?

- Who will be responsible for acquiring copyrights and licensing agreements?

- How will the content be archived and documented?

In chapters 3 and 4, you learned about the elements that make up multimedia titles: text, graphics, sound, animation, and video. These chapters also discuss the sources related to these elements: libraries of clip art, sound, and video; scanned photos and slides; draw and paint programs used to create graphics; video capture cards; MIDI synthesizers for music; and word processing programs for text. If the multimedia title repurposes existing content, such as a book or catalog, obtaining the material might be relatively easy. If original content must be created, however, especially animation and video, the process is more involved and often requires contracting with outside suppliers. If the goal is to market the title, quality is extremely important. Graphic artists and photographers would be contracted to create original artwork and pictures; actors would be employed for video production and narration; musicians might be hired to write and produce original music and sound effects; and editors would be used to review any text. If the subject matter was particularly technical or specialized, a content expert such as a teacher might be enlisted to create test questions.

Because a multimedia title can contain hundreds of images, it is important to provide for cataloging the various graphical elements. Figure 6.9 shows a

Figure 6.9

A database of images

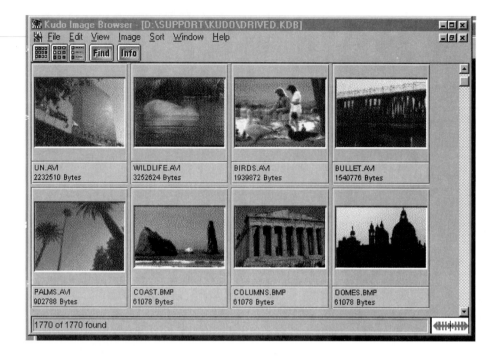

typical database of images that includes the name, type, and size of the image. This database would be used to keep track of the elements, determine total file size, and provide a reference for filenames that might be used in scripting. In addition, the original artwork, photographs, and other content should be archived to protect it from loss or destruction if something happened to the working copy.

STEP 8: AUTHORING THE TITLE

"Bringing it all together"

If the planning phase, especially developing the specifications and storyboard, has been completed properly, the authoring requirements would be fairly straightforward. This does not mean that the authoring would be easily or quickly accomplished, but that there would be a clear indication of what needs to be done. In determining which authoring system to use, a developer would consider the following points:

- The playback system (Macintosh and/or Windows-based computers)—some authoring programs work with only one platform

- The emphasis placed on animation—certain programs have fairly sophisticated 2-D animation tools

- The expertise of the programmer—programs differ greatly in the approach they use (metaphor) and their scripting language

In chapter 5 you learned that authoring a multimedia title can be as simple as creating an electronic slide show using presentation software such as Compel. You could select a background, type the text, scan some graphics, and import clip art, video, and sound. You could even create buttons with hyperlinks. The entire process would be inexpensive, relatively quick to complete, and require little specialized expertise. These types of titles might be appropriate for in-house training, lectures, conference presentations, sales presentations, and other noncommercial applications. If the goal is to create a commercial-quality stand-alone title, however, authoring plays a significant role. Invariably, scripting becomes a focal point in order to provide the functionality called for in the specifications. Scripting is needed for such things as checking user input from the keyboard, accessing an external device such as a videodisc, determining the configuration of the playback system, and creating an installation program that is used to start the title. It is critical that the programmer work closely with those designing the user interface and those providing the content to help ensure that the specifications are being met.

The Testing Phase

STEP 9: TESTING THE TITLE

"Does it work the way it was planned?"

Although it is listed as the final step in the development process, **testing** should be ongoing. In fact, testing can start at the very beginning during the concept stage. Small groups (focus groups) of potential users could be shown a prototype of the proposed title. The prototype could be as simple as an electronic slide show presentation with enough content (even rough graphics) and interactivity to demonstrate the concept and determine its feasibility. Conducting this "proof of concept" can provide valuable information that helps evolve the initial idea and prevent costly oversights.

Throughout the creation phase of the title, it is important to test the design and the function. Testing the design involves how the user interacts with the title and asks questions such as, Does the user understand the navigation scheme, terminology, icons, and metaphors? Does he or she get stuck, confused, or lose interest? Usability testing is a formal process in which potential users are videotaped as they interact with a title and asked to verbalize what they are thinking. This allows the developer to see what the users do (which objects are clicked and when), why they interact the way they do, and what their feelings are as they progress through the title.

Testing the function of a multimedia title involves making sure it works according to the specifications and answers questions such as, Does clicking on each button or object cause the appropriate action? Can animation, sound, and video clips be controlled by the user? Do graphics, text, and other elements appear in the correct locations? Although testing the functionality of the title continues throughout the creating phase, there are two formal processes: alpha testing and beta testing. ***Alpha testing*** is usually conducted in-house and is not restricted to the development team. The idea is to "try to make it crash," and every conceivable action (point and click) and navigation path should be explored.

Beta testing is the final functional test before release. It involves selected potential users that could number in the thousands—Windows 95 had 40,000 beta testers. Because the users are outside the company, they are often required to sign nondisclosure agreements (see figure 6.10) to prevent them from revealing information about the title before it is released. Companies try to make it easy for beta testers to provide ***feedback*** by giving them an e-mail address to contact and/or a disk containing a questionnaire that each tester fills out and returns in a prepaid package. To encourage a high level of participation, companies often provide an incentive such as discounts on the finished product. A goal of beta testing is to get feedback from as wide a variety of potential users as possible. Another goal is to have the test done on as many different computer configurations as possible. This is important, especially with Windows-based computers that can be assembled with a variety of components made by different companies.

To extend what you've learned, log on to the Internet at
http://www.thomson.com/wadsworth/shuman
You will find a wide variety of resources and activities related to this chapter.

Figure 6.10

A nondisclosure agreement

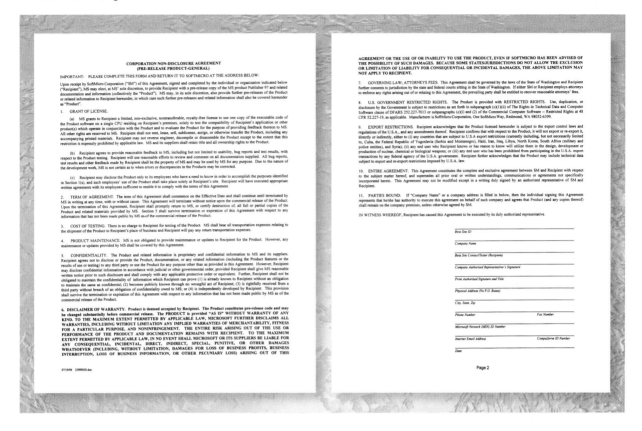

 key terms

- alpha testing
- beta testing
- demographics
- exploratory navigation
- feedback
- goals
- "look and feel"

- objectives
- sequential navigation
- specifications
- storyboard
- testing
- topical navigation
- user interface

review questions

1. The rule-of-thumb for multimedia development is _____ percent planning and _____ percent production.

2. **T F** Testing is an important step in multimedia development and should be done only at the end of the process.

3. One of the tenets of marketing is to find a need and _____ it.

4. Every multimedia project originates as an _____.

5. **T F** Goals are broad statements of what the project is about, whereas objectives are more precise and measurable statements of what a project will accomplish.

6. **T F** The "look and feel" of a multimedia title includes determining the tone.

7. **T F** The goal in creating the specifications is to be able to give them to a production team and, with little further instructions, have the team create the title.

8. The user interface involves the _____ and the _____.

9. A storyboard includes the exact content to be included in the multimedia title.

10. **T F** A topical navigation scheme takes the user through a controlled, linear process.

projects

Select a topic for a multimedia title and complete a report with the following:

1. Develop the concept and specify, in general, what you want to do.

2. State the purpose of the title, including the goals and objectives.

3. Identify the target audience, including demographics and lifestyle information.

4. Specify the treatment—the "look and feel." Include metaphors and emphasis you will place on various multimedia elements. Support your decision based on the target audience, concept, and objectives.

5. List the specifications, including:
   - Target playback system(s)
   - Elements to be included
   - Functionality
   - User interface

6. Design the storyboard and the navigation scheme. Include a template design if appropriate and specify the hyperlinks.

7. Indicate how and where the content will be acquired.

8. Specify when in the development process testing will occur and how it will be done.

9. Prepare a brief oral presentation of your report and be ready to present it to your class.

Designing for Multimedia

AFTER COMPLETING THIS CHAPTER YOU WILL BE ABLE TO:

- Describe how basic design principles can be used in designing multimedia titles
- Describe how the intended audience, type of title, and contents affect the interactive design
- Specify the guidelines for interactive design

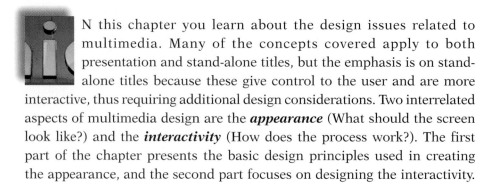I N this chapter you learn about the design issues related to multimedia. Many of the concepts covered apply to both presentation and stand-alone titles, but the emphasis is on stand-alone titles because these give control to the user and are more interactive, thus requiring additional design considerations. Two interrelated aspects of multimedia design are the ***appearance*** (What should the screen look like?) and the ***interactivity*** (How does the process work?). The first part of the chapter presents the basic design principles used in creating the appearance, and the second part focuses on designing the interactivity.

Creating the Appearance— Basic Design Principles

The application of ***design principles*** is important when determining how a screen will look in a multimedia title. Figure 7.1 shows a screen layout that is typical of many multimedia titles, especially corporate communications, reference, and training titles. The objects (heading, text, graphic, and navigation buttons) on the screen are fairly common to these titles, and the question becomes how to arrange them on the screen. The designer will try to maintain consistency in the "look and feel" of each screen even when the objects change. For example, figure 7.2 shows the same basic layout with a window to view a video clip where the text was previously located. Notice that the heading and navigation buttons are in the same place as in figure 7.1. Following are selected design principles and examples of their use in multimedia. These principles are presented as guidelines rather than absolute rules. Their application to any particular multimedia title, explored further later in this chapter, is based on the intended audience, content, and type of title. The overriding consideration is how to design the appearance of the screen to achieve the objectives of the title—to facilitate communication, provide entertainment, elicit emotion, and so on.

BALANCE

In general, as human beings we seek to maintain balance. We want equilibrium in our lives (balance work and play) and in our society (balance the scales of justice). Balance has a positive connotation, and, therefore, we may respond differently to visual images that are in balance and those that are out of balance. When we view things that are out of balance, we tend to feel that something is not quite right.

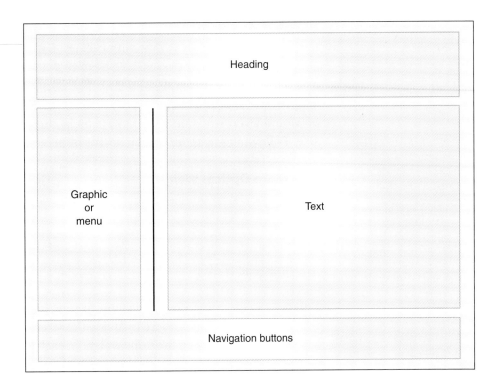

Figure 7.1
A screen layout typical of many multimedia titles

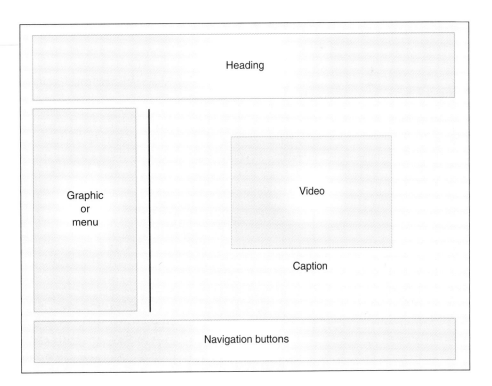

Figure 7.2
The same screen layout with a window to view a video clip

Balance in screen design refers to the distribution of optical weight in the layout. **Optical weight** is the ability of an element (graphic, text, headline, subhead, and so on) to attract the user's eye. Each element has optical weight as determined by its nature and size. The **nature** of an element refers to its shape, color, brightness, and type. For example, a stunning color photograph of Mount Everest would have more weight than a block of text of an equal size. Following are some guidelines for understanding optical weight:

More Optical Weight	Less Optical Weight
Large	Small
Dark	Light
Color	Black, white, gray
Irregular shape	Regular shape

Balance is determined by the weight of the elements and their position on the screen. That is, if you were to divide the screen into four parts, a balanced layout would have about the same weight in each part (see figure 7.3). Balance can be accomplished through symmetrical design or asymmetrical design. **Symmetrical** balance is achieved through arranging similar elements, such as two graphics of equal weight as shown in figure 7.3. **Asymmetrical** balance is achieved by arranging dissimilar elements as shown in figure 7.4. In general,

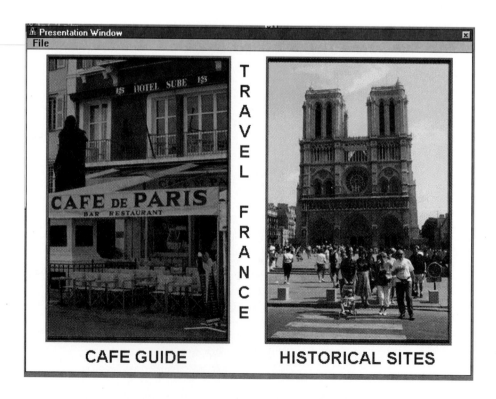

Figure 7.3

A balanced layout using a symmetrical design

Figure 7.4

A balanced layout using an asymmetrical design

symmetrical design is static and suggests order and formality. Symmetrical design might be appropriate for multimedia titles that focus on corporate image for conservative organizations such as banks and insurance companies. On the other hand, asymmetrical design is dynamic and suggests diversity and informality. It might be appropriate for entertainment titles, where a feeling of movement and discovery are important.

Following are ways to achieve balance in design:

- Use even numbers of elements of equal weight (symmetrical)
- Use two or more elements of smaller weight to balance one large element (asymmetrical)
- Enclose text in a box or use a color background to give it more weight
- Surround a dark graphic with abundant white space

Balance can be affected by assigning attributes to elements:

- **Size** Larger objects are perceived to be more important.
- **Position** Objects that are placed higher on the screen are thought of as more dominant.

- **Color** The color of an object can imply significance, such as red connoting heat, anger, or stop.

- **Space** Too many objects too close together may give the impression of clutter and disorganization, whereas a few objects with abundant white space could convey the opposite.

UNITY

Unity has to do with how the various screen elements relate—how they "fit in." Figure 7.5 shows a design in which each element complements the others; figure 7.6 shows an example of one element that is unrelated. An element that seems out of place can be disconcerting to the user and distract from achieving the desired effect of a particular screen. Unity reinforces the message or theme on individual screens (intrascreen unity) and provides consistency throughout the title (interscreen unity). In most cases, the goal is to have the elements complement each other. Unity can be achieved by consistency in shapes, colors, text styles, and themes.

In multimedia titles unity applies to the interactive design as users navigate from one screen to another (as discussed later in this chapter). Unity also applies to the design of each screen and to the design of parts of a screen. For example, you would want to relate headings with text, graphics with captions, and video clips with controls (see figure 7.7). Unity is a desirable

Figure 7.5

A design in which each element complements the others

design goal in many multimedia titles. In games and other entertainment titles, however, where exploration and surprise are important considerations, a unified design may prove rather dull.

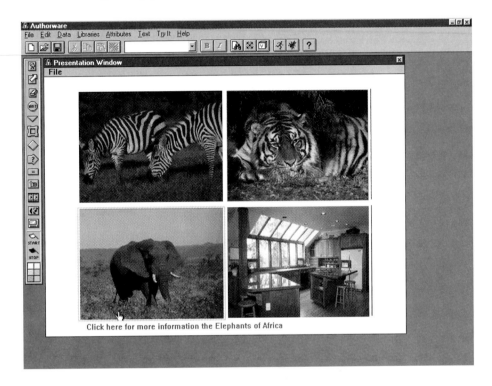

Figure 7.6

A design in which one element does not relate to the others

Figure 7.7

An example of unity in the design

MOVEMENT

Movement has to do with how the user works through the elements on the screen. When a screen appears on the monitor, the viewer's eye is drawn to a particular location. In a balanced design, this might be what is called the ***optical center***—a point somewhat above the physical center of the screen. The tendency is to move from upper left to lower right as we proceed though the contents of a screen. This is a phenomenon that causes many designers to put the Continue or Next button at the lower-right corner of the screen.

Movement is especially important in training and educational titles in which the designer wants the user to work through the contents in a more structured way. It is also important in situations where there is a primary message or impression that the designer wants to convey. In these cases, the designer will try to effect movement and emphasize various elements by applying certain design techniques:

- Controlling where the user starts on the screen by placing emphasis on a graphic, headline, or text block
- Creating asymmetrical balance
- Using lines or objects that point in a certain direction
- Using color gradients that go from a light shade to a dark shade
- Having people or animals looking in the direction you want the user to look (see figure 7.8)

The designer can emphasize an element by making it a contrasting shape or color, surrounding it with white space, using a different font or type style, creating borders, and using different backgrounds for selected objects.

interactive exercise

On the *Multimedia in Action* CD are several examples of basic design principles. Take a few moments to view these.

1. Start the CD.
2. Choose Design from the contents screen and read the instructions.
3. Click on the questions button and review the questions for Design Principles.
4. Click on the demos button and choose Design Principles.
5. After viewing the demo, respond to the questions previously reviewed.

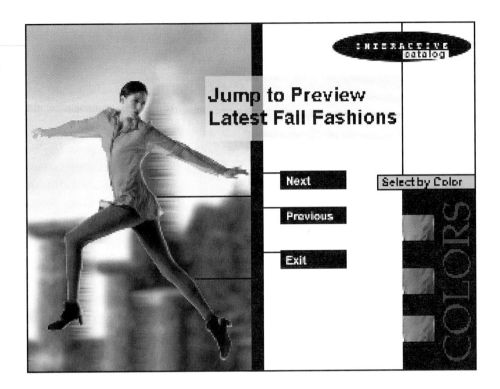

Figure 7.8

An example of movement in the design

Designing for Interactivity

The basic design principles presented here are, in varying degrees, useful for any multimedia title, but they focus on only one part of multimedia design—the appearance. When interactivity is built into the multimedia title, control is given to the user. The goal of multimedia design is to develop an interface that will allow user control in a way that works with the content while addressing the needs of the user (for example, if a sound is played, the user should be able to adjust the volume). In other words, the design needs to be user-centered. In this section you learn about design issues focused more on interactivity, that is, how the user works through the multimedia experience. Keep in mind that designing the appearance and designing the interactivity are interrelated and must complement one other. Several things affect the nature of the interactive design, including the audience, type of title, and content.

AUDIENCE

Foremost in guiding the interactive design process is the end user. As much as possible, the designer must understand the users' needs and how they work with the product. The tendency of the designer is to approach the design process from his or her own perspective, reflecting personal knowledge and experiences. These, of course, can be quite different from those of the intended audience. Designers might be computer "power users," understanding graphical interfaces, metaphors, icons, and menu-based navigational schemes, whereas the intended user might be a novice who is intimidated by the computer. Even if users are familiar with computers, they may make certain assumptions about how the interaction works, based on their experience. For example, they might double-click the mouse button to open a file folder displayed on the screen, because this is how their word processing program works. This user action might not have occurred to the multimedia designer, who specifies that the object (file folder) should respond to a single click. The challenge for designers is to put themselves in the place of the user by asking some very fundamental questions:

- What does the user see on the screen?
- What does the user want to do?
- What is the user's realm of experience?

TYPE OF TITLE

The type of title affects the design as illustrated in the following examples.

- If the title is a corporate annual report and the user wants to quickly access specific information, the design might need to accommodate only a main menu with a straightforward navigation scheme and the user's ability to point and click the mouse (see figure 7.9).

- If the title is a reference work, such as an encyclopedia, the design might become more complex as the user is given the ability to access a topic in several ways—menu, index, object, or key word search (see figure 7.10). With a search function, keyboard input or a simulated on-screen keyboard that could be used with the mouse pointer would need to be provided. If keyboard input is allowed, how the user works with the keyboard must be considered. For example, if the instructions are to "Press the Return key," a Macintosh user would understand the instructions; but because there is no Return key (but rather an Enter key) on the PC keyboard, a Windows user might not intuitively understand the instructions.

Chapter 7 Designing for Multimedia

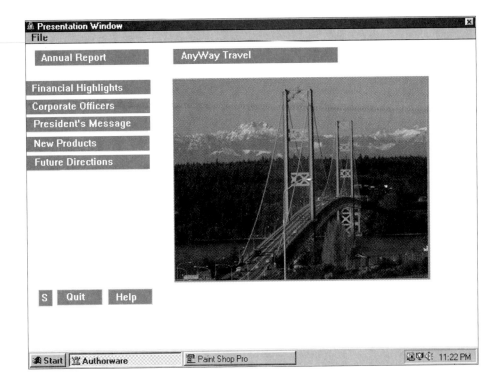

Figure 7.9

A navigation scheme designed to allow quick access to specific information

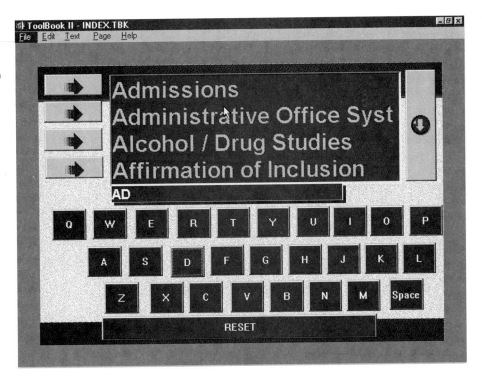

Figure 7.10

A navigation scheme that allows for a key word search

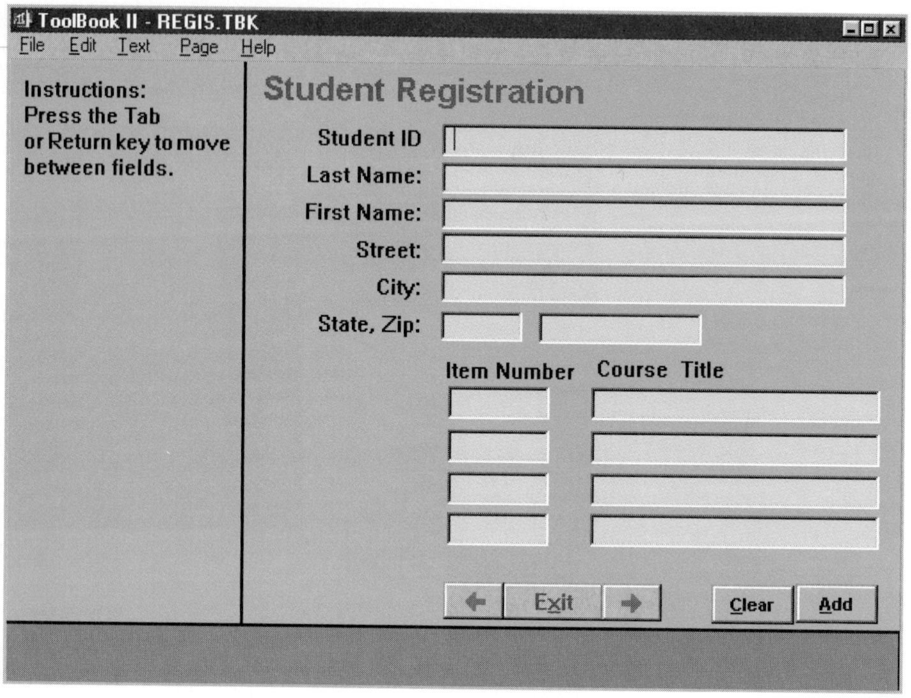

Figure 7.11

A title that allows for transactions

- If the title allows transactions, such as a registration kiosk at a college (see figure 7.11), the designer must consider how the users know which field (box) is active, how to move from field to field, and how to indicate when they are done with a field and when they are done with the form.

- If the title is a game and the focus is on entertainment, the design might allow random interaction. Because the focus is on exploration and discovery, users are not as concerned with where they are and what they can do next. Thus the user interface becomes more implicit (with invisible hot spots) rather than explicit (with visible buttons).

CONTENT

Following are examples of the ways that content can influence the design of a title's interactivity.

Large amounts of content   One of the major considerations in designing interactivity is how many levels to navigate. The tendency is to add more levels as the content increases. The more levels, however, the greater the chance for confusion and frustration as users try to determine where they

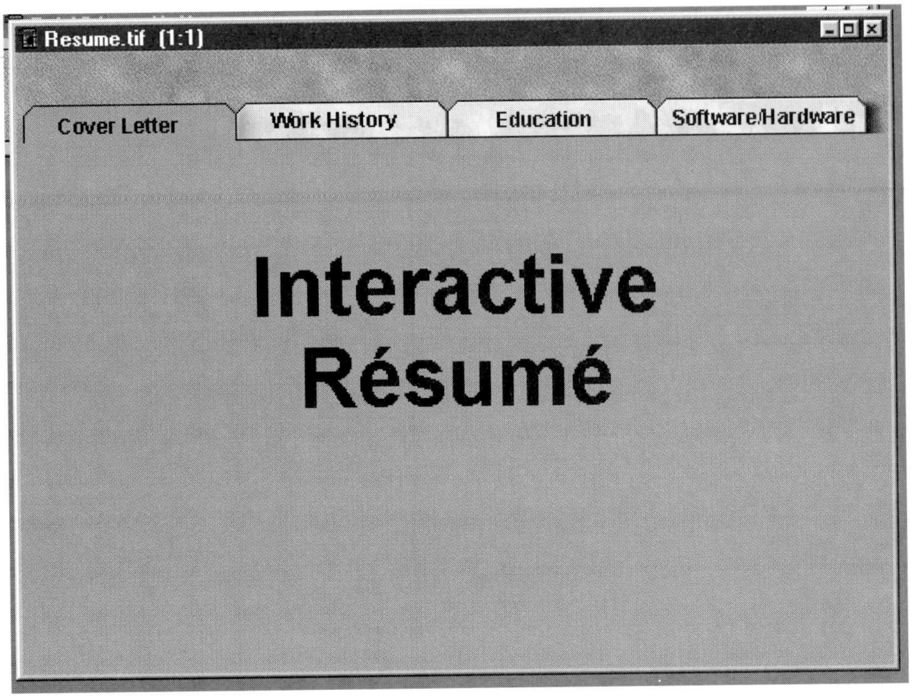

Figure 7.12

A title using tabs

[Screenshot showing a window titled "Resume.tif (1:1)" with four tabs: Cover Letter, Work History, Education, Software/Hardware. The main content area displays "Interactive Résumé".]

are, how they got there, and how they can get back to where they started. There are numerous ways to reduce the levels in a multimedia title:

- Provide shortcuts in the form of hotwords or hot spots that skip several levels
- Replace parts of the original screen with new content, but leave the shell of the screen intact to maintain the user's frame of reference
- Use pop-up windows that display additional information
- Use scroll bars for text-intensive titles
- Provide tabs or bookmarks that indicate where the user has been and allow the user to quickly return to a previously viewed screen (see figure 7.12)

Elements used in the title Another way content affects interactive design is in the elements used. As previously mentioned there are ways to accommodate text-intensive titles. If video (or animation or sound) is used, there are certain decisions to be made that affect interactive design:

- Who controls the video? Does the video play automatically or can the user stop and start it?

Figure 7.13

Typical VCR controls used in a title

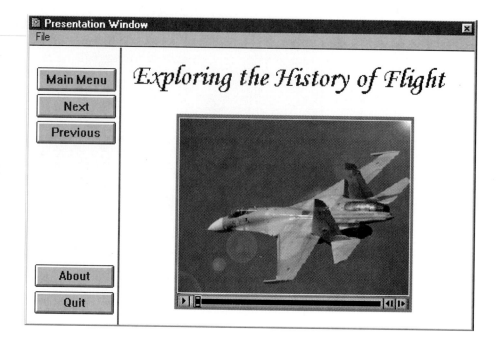

- If it starts automatically, can the user pause, stop, rewind, and cancel it? The user may have previously viewed the video and does not want to see it again.

- What will the controls look like? Most people are familiar with standard VCR controls as shown in figure 7.13, but these controls may not visually complement the rest of the screen. Figure 7.14 shows controls that fit the theme of the title.

- Can the sound volume be adjusted?

- Where and when does the video image appear on the screen?

Nature of the content The content can suggest a theme that the interactive design would need to complement and reinforce, as illustrated in the following examples:

- In a title on astronomy in which students study the night sky, the view could be through a telescope that the user manipulates with the mouse.

- A reference title on the history of cinema might include a control room with buttons that are used to open curtains and play video clips that appear in a window resembling a movie screen.

Figure 7.14

Controls that fit the theme of the title

- In a title that is a walk-through of a museum, the interactive design might include a virtual reality environment in which the mouse pointer is used to navigate through the museum and the mouse button is used to zoom in on paintings (see figure 7.15).

Figure 7.15

An example of an interactive design that reinforces the theme of the title

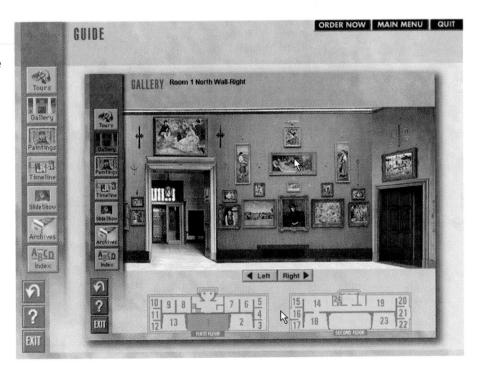

Guidelines for Interactive Design

Following are general guidelines that are useful in designing interactive multimedia, especially informational titles.

MAKE IT SIMPLE, EASY TO UNDERSTAND, AND EASY TO USE

The entire design process is for the benefit of the user. The user should not have to be taught how the navigation scheme and media controls work; they should be intuitive. From the moment the first screen appears and throughout the interactive process, users should know where they are and where they can go (unless the title is designed to be exploratory, such as a mystery or adventure game). The initial screen should provide an indication of what is contained in the title and how to navigate through it.

Figure 7.16 shows an example of the beginning screen suggesting the content and navigation scheme. Metaphors should be within the user's frame of reference and consistent with the content. All of the screen images including icons and pointer symbols as well as the navigation process should complement each other and be consistent with the title's theme.

Figure 7.16

An example of the initial screen suggesting the navigation scheme and content

BUILD IN CONSISTENCY

Consistency is especially important for reference titles in which the user is searching for specific information. Consistency applies to both the appearance of each screen and how the navigation scheme works. Figure 7.17 shows how a consistent look is maintained as the user navigates from one

Figure 7.17
How a consistent look is maintained as the user navigates from one screen to another

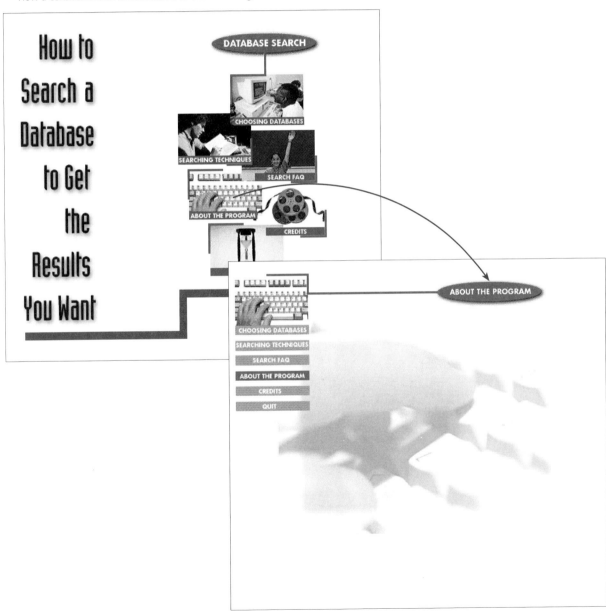

level to another. In this case, the graphic in the first screen that is used to navigate to the second screen becomes the background for the second screen. This not only provides unity among various screens, but helps the user understand the navigation process.

USE DESIGN TEMPLATES

A *template* is a precise layout indicating where various elements will appear on the screen (see figure 7.18). Templates can aid the design process in several ways:

- **Provide consistency** Each element of the screen will be in the same location, which aids the user in understanding how the title works and increases the speed at which the user can navigate through the title.

- **Shorten the development time** Given the similarity of many screens, templates can reduce the amount of time needed to arrange elements on the various screens.

- **Prevent "object shift"** An object that moves even one pixel as the user navigates through the title causes a noticeable and disconcerting jump. Templates that utilize grids can specify the exact layout, down to a pixel, of each screen element and prevent objects from shifting.

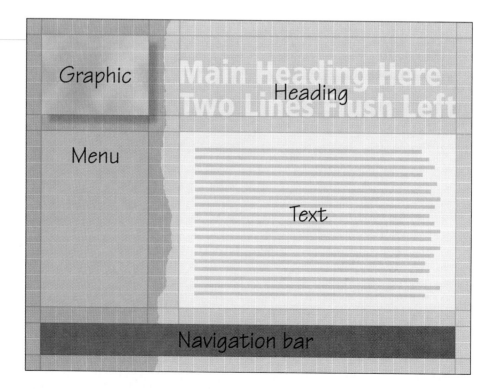

Figure 7.18

A design template

Chapter 7 Designing for Multimedia

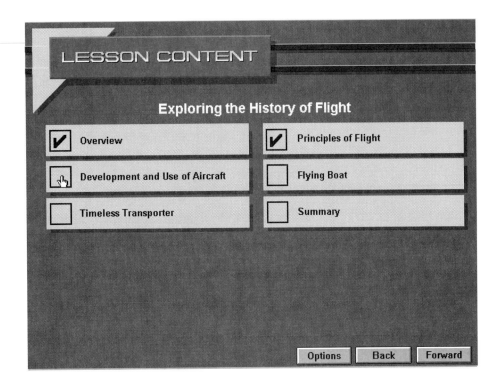

Figure 7.19
Providing feedback to the user

PROVIDE FEEDBACK

Interactivity is a two-way process. Users need to know when a button has been clicked—it should change to a depressed state, change color, or play a sound. Users also need to know when an object has been selected—it should appear highlighted, a different color, or surrounded with a border. Another form of feedback is shown in figure 7.19, where check marks are used to indicate where the user has been or what has already been viewed.

PROVIDE CHOICES AND ESCAPES

Avoid long introductions of automatically scrolling text, narration, music, or credits. Don't make the user view a video or animation, or listen all the way through sound clips that may already have been heard. Provide a way for the user to skip or escape from viewing these elements. Additionally, care should be taken in the use of a Quit button, especially if it takes the user completely out of the program. To avoid accidentally quitting, a prompt should appear when the Quit button is selected to allow the user to confirm the desire to quit.

interactive exercise

On the *Multimedia in Action* CD are examples of various interactive designs. Take a few moments to view these.

1. Start the CD.
2. Choose Design from the contents screen and read the instructions.
3. Click on the questions button and review the questions for Interactive Designs.
4. Click on the demos button and choose Interactive Designs.
5. After viewing the demo, respond to the questions previously reviewed.

Internet Link

To extend what you've learned, log on to the Internet at

http://www.thomson.com/wadsworth/shuman

You will find a wide variety of resources and activities related to this chapter.

key terms

- appearance
- asymmetrical
- balance
- design principles
- interactivity
- movement
- nature
- optical center
- optical weight
- symmetrical
- template
- unity

review questions

1. **T F** Balance is determined by the weight of elements and their position on the screen.

2. _____ balance is achieved through arranging similar elements.

3. **T F** Graphics and text (but not headings) are considered elements when determining balance.

4. Unity is achieved by _____ in shape, color, text styles, and theme.

5. **T F** A regular shape has a higher optical weight than an irregular shape.

6. **T F** Unity applies to both interscreen and intrascreen design.

7. **T F** The optical center is the exact physical center of the screen.

8. **T F** Movement in design has to do with the frame rate of an animation.

9. In a balanced design, the viewer's eye may be drawn to a point on the screen called the _____.

10. **T F** Using lines or objects that point in a certain direction is a way to effect movement.

11. **T F** Generally, the design of a multimedia title should provide a way for a user to escape from animation or video clips that have already been viewed.

projects

1. Select several multimedia titles and study their screen designs on the basis of appearance. Prepare a report including the following:

   - For a screen that has symmetrical balance:

     How are the elements (heading, text, graphics, colors) used to achieve the balance?

     How does this design enhance the title?

   - For a screen that has asymmetrical balance:

     How are the elements (heading, text, graphics, colors) used to achieve the balance?

     How does this design enhance the title?

   - For a screen that has obvious movement:

     How is the movement achieved?

     How does the movement enhance the design?

   - For a screen with an obvious optical center:

     What is the optical center?

     How was the screen designed to achieve the optical center?

     What purpose does the optical center achieve?

   - What changes would you make to improve the screens you studied?

   Prepare a brief oral presentation of your report and be ready to present it to your class.

2. Select several multimedia titles and study their screen designs on the basis of interactivity. Prepare a report including the following:

   - How has the interactivity been designed to address the needs of the intended audience?
   - How has the interactivity been designed to address the type of title?
   - How has the interactivity been designed to address the content?
   - Is the design of the interactivity intuitive and easy to understand? Explain.
   - Is the design of the interactivity consistent? Explain.
   - Does the design of the interactivity provide needed feedback and choices for the user? Explain.
   - What changes would you make to improve the screens you studied?

   Prepare a brief oral presentation of your report and be ready to present it to your class.

8

Managing Multimedia Projects

AFTER COMPLETING THIS CHAPTER YOU WILL BE ABLE TO:

- Describe the management issues involved in developing multimedia
- Discuss the major management areas related to multimedia projects
- Describe the various specialists in multimedia
- Indicate the role of budgets and schedules in managing multimedia projects

Management Issues of Multimedia Development

In this chapter you learn about the management process and the management skills related to developing multimedia titles. Managing the development of multimedia titles is basically project management with the focus on such items as objectives, budgets, and timelines. The nature of multimedia development, however, places the emphasis on people. In a manufacturing process where the goal is to produce a tangible product, managers are concerned with, among other things, ordering parts, inventory control, packaging, and shipping. The process of developing multimedia titles does not produce a tangible product per se. Although the multimedia title can be distributed using CDs, kiosks, and the Internet, the essence of the "product" is its content assembled as an application (a game, training tool, reference source, or sales aid, for example) that runs on a computer and which has been created by highly skilled professionals. Thus the management focus is on people—their creative talents, experience, and expertise. As one corporate executive said, "Every evening our inventory goes down the elevator."

WHO WILL MANAGE THE DEVELOPMENT PROCESS?

After making the decision to create a multimedia title, a company needs to determine who will manage the project—will it be done in-house or by contracting out to a service provider. If the company's business is to make multimedia titles, most likely the decision will be to manage the development process in-house. For non-multimedia companies, however, the decision may be more involved. Following are three categories of organizations and how they might approach the question of who manages the multimedia development process.

Non-multimedia company Many organizations are interested in using multimedia for presentations, corporate training, trade show displays, product catalogs, and sales kiosks. Yet multimedia is not the company's business. The company must therefore make the decision of producing the title in-house or contracting it out to a multimedia service provider, a company that specializes in developing multimedia. The decision is affected by the following considerations.

- **Complexity of project** A simple electronic slide show presentation might easily be produced in-house, whereas an interactive corporate image title including video, sound, and animation would be beyond the scope of the typical company.

- **Expertise of current staff** Some large companies, such as AT&T and Boeing, have corporate divisions that produce multimedia training titles. In most cases, however, companies do not have on staff the highly specialized talent that is needed to develop multimedia titles and manage the development process. Thus the decision becomes whether or not to train the current staff, hire the needed personnel, or contract with a multimedia service provider.

Training existing personnel may be appropriate if a company determines that there is an ongoing development need. The expense could be considered an investment that would have a payback for years to come. There may also be talented individuals whose skills are readily transferrable to digital technology. For example: Graphic artists could learn interface design and be trained to use image processing programs such as Photoshop; programmers could be trained on authoring programs such as Director or ToolBook; and technical writers could learn how to edit content for multimedia titles.

Even if a company makes a commitment to establish a comprehensive multimedia development division, there will always be the need, depending on the nature of the project, for specialized talent. It is rare that a company would be able to support a staff that includes such highly specialized skills as sound engineers, video producers, and 3-D animators. Thus, often a company that decides to create a multimedia division will contract out the most specialized tasks. There are agencies devoted specifically to matching a company's needs with available talent, much like an employment agency. The difference between these agencies and others (such as temp agencies that provide employees to fill in for short periods of time and career agencies that provide permanent employees) is that they provide freelance employees who contract for a specific project. Contracting fits extremely well with the nature of multimedia projects, which are typically several months in duration. These agencies relieve the company of the expense of recruiting and the employee of extensive job hunting. In fact, many employees prefer this arrangement, because they are not required to "take the work home with them" as might be the case with a permanent position. In addition, they can receive benefits (medical insurance) and are exposed to a wide variety of experiences that can broaden their background and contribute to their portfolio.

Most companies do not have the expertise to develop multimedia titles in-house, either by using current employees or contracting for freelance help.

Therefore it is likely that a company will decide to contract out the entire development process to a comprehensive service provider.

Multimedia service provider A multimedia *service provider* can essentially become the multimedia arm of a company and take over the primary function of managing the project. In this case, it is important that the relationship be approached as a partnership, because the success of the project depends on close cooperation and effective communication. For example, even though the "client" is relied upon to provide the idea, objectives, specifications, and content, the service provider must understand exactly what the client needs. This is often difficult, because clients unfamiliar with multimedia may not have their needs clearly defined. Following is a list of project responsibilities for the client and the service provider.

Client	**Service Provider**
Providing idea	Helping refine idea
Stating objectives	Helping clarify objectives
Developing specifications	Clarifying and meeting specifications
Providing content as appropriate	Generating content as appropriate
Providing the budget	Providing cost estimates
Determining initial timeline	Establishing schedules and meeting timelines
Identifying company liaison	Identifying company liaison
Identifying the decision maker	Project management

Although each of the responsibilities is important, the two primary considerations are agreement on the specifications and consensus on who makes the final decisions. Specifications to a multimedia developer are like blueprints to a building contractor: They help in determining not only the tasks that are required, but also which subcontractors are needed, the project costs, and the schedule. Changes in the specifications often result in additional costs. For example, during the development process, product testing might indicate that users prefer to have the ability to adjust the volume and turn off a segment of audio. Adding this feature would require changes to the screen design so that a control (audio on/off and volume adjustment) is available. A related consideration is who in the client company makes the final decisions concerning the project. Often a client company may have several individuals involved or interested in the project (for example, public relations, marketing, and top management). These people may not even appear until after the project is under way. The goal is to avoid a situation in which substantial work has been done and then a "new player" (such as a corporate executive in another state) sees the title for the first time and suggests significant changes. It is important that the client determine who is involved in the decision making and keep them "in the loop" from the beginning.

Service providers know the industry. They can choose the appropriate specialists based on the client's needs. For example, creating a corporate image multimedia presentation for a bank would certainly take a different approach with a different "look and feel" than one for a rock band. The service provider would know which freelance interface designer would be most appropriate for each approach. The criteria used to evaluate service providers include their track record, as demonstrated by viewing past projects or provided through testimonials, references, and portfolios. Figure 8.1 shows an excerpt from a multimedia provider's promotional packet that lists their philosophy and services. It is important to note that this company emphasizes a long-term relationship with the client as their goal. Figure 8.2 shows an excerpt from a typical contract, and figure 8.3 shows a rate schedule for development work.

Multimedia company Companies such as Living Books, Brøderbund, Microsoft, Claris, and Sierra On-Line are in the business of creating multimedia titles for the education, home, and business markets. They produce and distribute children's and adult games, reference works, and training titles and exercise complete control over the management of their projects. Even so, these companies may find it necessary or advantageous to hire freelance help for specific projects or contract for specialized services such as video production that requires a studio.

The Management Process and Multimedia Projects

No matter who controls the management of a multimedia project (in-house, service provider, or multimedia company), there are certain functions that must be fulfilled. This section discusses the three primary functions related to managing multimedia projects: planning the project, organizing the resources and forming the team, and leading the team.

Figure 8.1

An excerpt from a multimedia provider's promotional packet

mdc
multimedia design center

PHILOSOPHY

MDC approaches every relationship with the assumption that it will be a long-term one. We feel that the client/provider relationship is as important as the end product. We understand that forming partnerships is the best way to ensure successful projects and believe that investing in the relationship pays returns.

COMMITMENT

Development of effective multimedia requires, among other things, two critical elements: (1) creative and technically astute developers and (2) the ability to keep up in an incredibly fast-changing environment. We are proud of our talented employees and are dedicated to staying current in the ever-changing field of multimedia in hardware, software, and staff expertise.

EXPERIENCE

MDC has been developing multimedia for more than 10 years. Originally founded by two former employees of Miconics, Inc., as a presentation development firm, we quickly grew to a full-service multimedia provider. Clients include First Mutual Bank, Western Airlines, Thunderbird Hotels, Peak Software, and the Heritage Museum.

SERVICES

MDC provides comprehensive multimedia project design, development, and implementation, including:

- **Corporate training**—interactive courseware and course management systems
- **Marketing materials**—product catalogs, brochures, and demonstrations
- **Interactive presentations**—for small or large groups, meetings, and conferences
- **Kiosks**—informational or transaction-based

We develop for all types of delivery media (CD-ROM, Internet, videodisc, CD-I, and so forth) and both Macintosh and Windows-based computers.

One Main Street ■ Scottsdale, AZ 85001 ■ (602) 555-1234

Figure 8.2

An excerpt from a multimedia provider's contract

mdc
multimedia design center

CONTRACT FOR SERVICES

Between: _____ (referred to as "client")

and: **Multimedia Design Center** (referred to as "provider")

The parties agree as follows:

1. **SERVICES** The provider shall perform services for the client with the attached schedule.

2. **PAYMENT** The client shall pay the provider according to the attached payment schedule for performing the required services.

3. **CONFLICT** The provider shall not perform any services for any person during the term of the contract where the performance of those services may or does result in a conflict of interest between the obligations of the provider to the client and the obligations of the provider to that other person.

4. **PROPERTY RIGHTS** Any information, data, programs, and products provided by the client to the provider, or developed while performing this contract, are the property of the client. The provider shall return all information to the client at the end of the contract.

5. **CONFIDENTIALITY** The provider shall keep the information and the contract in strict confidence unless it is in the public domain or the client has authorized disclosure.

6. **TERMINATION** The client may terminate this contract for any reason by giving the provider written notice of not less than 30 days.

7. **DISPUTES** All disputes arising out of or in connection with this contract or in respect of any defined legal relationship associated with it shall be referred to and finally resolved by arbitration under the rules of the America Arbitration Society.

8. **LAW** The laws of the State of Arizona govern this contract, and the provider shall perform the services in accordance with all applicable laws.

Client _____ Date _____

Provider _____ Date _____

One Main Street ■ Scottsdale, AZ 85001 ■ (602) 555-1234

Figure 8.3
A rate card from a multimedia provider

mdc
multimedia design center

RATE CARD

Multimedia project development .... negotiated

CD-ROM mastering .... bid

Professional services $120/hour

    Consulting (strategic planning, technical assistance)

    Production (graphics, animation, video, sound)

    Web site development

Training .... call for rates and schedule

    Regularly scheduled classes in fully equipped lab setting for popular development tools (Photoshop, Director, 3-D Studio, etc.)

    Sessions on multimedia and Internet strategies

    Custom sessions (one-to-one) or small groups

Workstation access

    Self-serve $50/hour

    With professional assistance $90/hour

One Main Street ■ Scottsdale, AZ 85001 ■ (602) 555-1234

PLANNING THE PROJECT

From a corporate standpoint, ***planning*** involves determining where you are and where you want to be. A company that creates children's educational titles may want to be the industry leader, which could be measured in such terms as sales, profitability, return on investment, awards, or market share. If a company determines that it has a 5 percent share of the market and the number one company has 25 percent, the first knows where it is and what it must do to be the leader. This type of corporate planning is focused more on long-range strategy than day-to-day operations.

From a project standpoint, planning involves how to achieve the objectives of an individual title. In chapter 6 you learned about the steps used in creating multimedia titles. Here you learn about planning as it relates to the management of an entire multimedia project. Three important planning areas are task analysis, budget, and schedule.

Task analysis   Based on the specifications, a determination could be made of the tasks necessary to create the desired title. The ***task analysis*** would determine the following:

- How much and what kind of video, audio, and animation is required?
- What content will be provided by the client and what will be created?
- How many illustrations? What type? How are they generated?
- What are the features that affect programming? User input? Access to databases? User testing?
- Will the title involve external media or hardware? Videodiscs, kiosks, printers?
- How many and what type of usability tests are planned?
- What are the licensing and copyright considerations?

The project planner uses this analysis to determine the required resources and the budget.

Budget   A ***budget*** is a critical part of project management because it provides the financial plan that affects a number of decisions. For example, the budget might determine if it is feasible to create original music, license commercially available music, or settle for sound clips. The budget also provides a control mechanism that is used to evaluate the status of a project. If during the project a comparison of the planned budget with the actual expenditures shows a discrepancy, a decision needs to be made on whether to adjust the original budget or reduce some planned expenditures. Figure 8.4 shows a budget for a multimedia title and a chart indicating how the funds

Figure 8.4

A budget for a multimedia title

mdc
multimedia design center

PROPOSED BUDGET

Project Title: **Museum Tour**

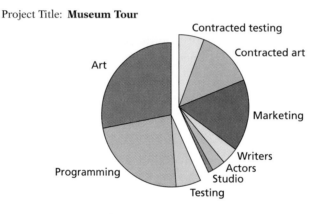

In-house development

Art	$ 180,000
Programming	$ 150,000
Testing	$ 40,000
Total estimated cost for in-house development	$ 370,000

Outside development

Studio	$ 6,000
Actors	$ 20,000
Writers	$ 25,000
Marketing	$ 100,000
Contracted art	$ 80,000
Contracted testing	$ 35,000
Total estimated cost for outside development	$ 266,000

Total proposed budget .... $ 636,000

One Main Street ■ Scottsdale, AZ 85001 ■ (602) 555-1234

are allocated. Although the costs of developing titles vary, depending on the content, features, and quality, it would be difficult to produce a commercially viable title for less than $100,000. More typical costs start at around $250,000.

Schedule A *schedule* specifies the project's beginning and ending dates as well as the milestones in project development. If a project misses the targeted completion date, the company may suffer in two ways. First, there will inevitably be cost overruns, because multimedia development is so people intensive, and the budget may be partly based on the number of person hours used. Second, missing the targeted completion date could cause a company to lose sales because of a delay in shipping the product. *Milestones* are significant accomplishments, such as the development of a prototype for usability testing, and are used to gauge progress in project development. Whether or not a milestone is achieved on schedule is critical to management control of the project.

A schedule is useful in coordinating the efforts of team members by showing how the various activities relate to one another. It can also indicate where potential bottlenecks might occur and where additional resources might be needed. Figure 8.5 shows a project schedule with various tasks and their planned beginning and ending dates. The schedule includes the sequence of activities and how they overlap, as well as what has been completed to date. A concern in the management of multimedia projects related to keeping on schedule is how to avoid so-called *feature creep*. This occurs in the development process when the specifications are changed by adding new features. This is a common phenomenon as companies try to enhance a product as it is being produced. Unfortunately, it often results in additional costs and missed deadlines.

ORGANIZING RESOURCES AND FORMING THE TEAM

Organizing involves making sure the necessary resources, especially personnel, are available. A critical aspect of multimedia development is the formation of the development team, which involves assessing the current talent in light of the requirements of a particular project. Whether shooting photos, scanning artwork, digitizing video, recording music, programming the interactivity, or testing the end product, people are the key. The challenge for managers is to bring together an often diverse group (such as designers with creative visions and programmers who must turn the visions into products) and keep them focused on the project objectives. Organizing resources addresses the following questions:

- Do we have the needed skills on our staff?

- Can we train current staff for skills we don't have?

Figure 8.5

A project schedule

mdc
multimedia design center

PROJECT SCHEDULE

Project Title: _____ Date: _____

Project Director: _____

ID	Task	Duration	Start Date	Finish Date
1	Refine concept and develop initial budget	1 week	2/1	2/7
2	Conduct planning session	4 days	2/7	2/10
3	Determine objectives, audience, treatment	5 days	2/10	2/14
4	Approve initial design	1 week	2/15	2/21
5	Develop specifications	2 weeks	2/21	3/6
6	Create storyboard	2 weeks	2/21	3/6
7	Approve final design	5 days	3/6	3/10
8	Approve final budget	5 days	3/10	3/14
9	Acquire content, create graphics and animations	3 months	3/10	6/10
10	Create prototype	1 month	3/10	4/10
11	Conduct feasibility testing	5 days	4/10	4/14
12	Programming	6 months	3/10	9/10
13	Conduct usability testing	1 week	8/1	8/7
14	Design package and label	2 weeks	6/15	6/28
15	Create graphics for packaging	1 month	6/29	7/29
16	Create premastered CD	5 days	9/10	9/14
17	CD mastering	15 days	9/16	9/30
18	Ship	0 days	10/1	

One Main Street ■ Scottsdale, AZ 85001 ■ (602) 555-1234

- What do we need to contract out?
- What are the specific assignments?
- What physical resources are needed (new or specialized computer equipment, memory, or storage capacity upgrades, new or specialized software or upgrades, slide scanner, digital camera)?
- What is the communication process (e-mail with shared folders, periodic team meetings, off-site planning sessions)?
- How is project content shared (company network, external storage devices such as Syquest or Zip drives)?

Forming the Team Multimedia development, especially for non-presentation titles, is not a one-person show. Because of the specialized skills required, a team approach is used. At minimum the team would need design and programming expertise as well as a content provider. Depending upon the type, features, and content of a title, the team could involve several specialists. Following is a description of selected job titles. (Note that the specific titles may vary.)

- **Graphic artist (illustrator, image editor)** Creates and manipulates graphic image
- **Content (subject matter) expert** A person considered knowledgeable about the subject matter who provides content (such as an author, editor, teacher, engineer, scholar, or other professional)
- **Programmer** Brings all the pieces together and makes it work, utilizing programming languages and/or authoring tools
- **Interface designer** Understands how users work with computers and how screen layout affects user interactivity
- **Project or program manager** Oversees the project, utilizing management skills and knowledge of the multimedia development process
- **Media professionals (photographer, videographer, audio engineer)** Create photos, video, and audio content
- **Usability tester** Designs and conducts product testing
- **Acquisitions specialist** Acquires content, especially involving licensing or copyright issues
- **Animator** Creates 2-D and 3-D animation
- **Instructional designer** Knows learning theory and can apply it to multimedia design

More than one of these functions can be fulfilled by a single individual. What matters most is determining which functions are needed and how to acquire the necessary talent.

LEADING THE TEAM

Leading involves influencing others to achieve the project goals. This is extremely critical in multimedia development, because each member of the team brings highly technical and specialized skills to the project. Team members' tasks are interdependent, and an unmotivated team member can be extremely detrimental to the project. The project manager must have good leadership skills and set an environment that fosters teamwork.

The following are some important characteristics of a good team leader:

- **Lead by example** Others will take their cue from the project leader, who will want to be viewed as part of the group instead of above the group.

- **Be straightforward and sincere** It is important to establish a sense of trust among the team members, which in turn leads to more open and effective communication.

- **Recognize accomplishments through praise and rewards** People are encouraged and motivated when they feel their contributions have been recognized.

- **Listen** Because of the diverse nature of a multimedia development team, each member provides a unique perspective that needs to be heard.

Following are some of the ways to establish an environment that fosters teamwork:

- If possible, set up the project so that the team has the task of creating the complete multimedia title. This helps them take ownership of the project and fosters pride in their work. Include a Credits screen (as an option for the user to view) that lists the team members and their roles.

- Involve the team as early as possible in the development process. Empower the team members by providing opportunities for them to give input and participate in decision making. Hold periodic meetings to review the project's status and encourage involvement by all the team members. Rotate some duties, such as the "facilitator" during team meetings, to promote the sense of equality.

- If possible, have the team members located near each other. Display the original storyboard in a prominent location to remind the team of the project's goals and status. Set up an e-mail folder for the team to facilitate communication—and encourage humor.

To extend what you've learned, log on to the Internet at
http://www.thomson.com/wadsworth/shuman
You will find a wide variety of resources and activities related to this chapter.

key terms

budget
feature creep
leading
milestones
organizing

planning
schedule
service provider
task analysis

review questions

1. In developing multimedia titles, the management focus in on _____.

2. **T F** Some agencies specialize in providing freelance employees who contract for a specific multimedia project.

3. **T F** Comprehensive multimedia service providers are used by a company to provide ideas for multimedia titles.

4. **T F** From a corporate standpoint, planning involves determining where you are and where you want to be.

5. Three important planning areas are:
   a. _____
   b. _____
   c. _____

6. If a project misses the target completion date, the company may suffer in two ways: cost _____ and lost _____.

7. _____ occurs when, during the development process, the specifications are changed by adding new features.

8. The job title of a person who knows learning theory and can apply it to multimedia design is: _____.

9. **T F** Leading involves influencing others to achieve the project goals.

10. The _____ is used to determine the required resources and the budget.

projects

1. Contact a comprehensive multimedia service provider and conduct the following research:
   - Determine what services are provided
   - Determine who their clients are
   - Determine who their major competitors are
   - Determine what their major accomplishments are
   - Determine how they acquire new clients
   - Determine how they advertise and promote themselves
   - Obtain a rate card
   - Obtain a sample contract
   - Obtain a promotional packet

   Prepare a brief oral presentation of your report and be ready to present it to your class.

2. Contact an employment agency that specializes in (or has provided) contract employees for multimedia-related positions within a company and determine the following:
   - What do they see as the benefits to the employee and the benefits to the company?
   - The employment agreement in terms of the responsibilities of the three parties (agency, employee, company)
   - Pay rate
   - Benefits package
   - Length of employment
   - Guarantees
   - Evaluation process
   - Other

   Prepare a brief oral presentation of your report and be ready to present it to your class.

3. Through library research, online research, and interviews, determine the employment trends in one multimedia-related position, such as those mentioned in this chapter. Develop a report that includes the following:

   - Position title
   - Position description
   - Trends in employment growth (Is the position forecast to increase or decrease in employment?)
   - Pay range
   - Skills needed
   - Educational background required

   Prepare a brief oral presentation of your report and be ready to present it to your class.

Part IV

Producing and Distributing Multimedia

9

Producing Multimedia Titles

AFTER COMPLETING THIS CHAPTER YOU WILL BE ABLE TO:

- Describe the issues involved in producing a title in-house

- Distinguish among the different CD formats

- Describe the process for producing CD-ROMs

PRODUCTION is concerned with preparing the multimedia title for use either as part of a presentation (sales, education, conferences), or for in-house training, kiosk delivery, or CD-ROM distribution. After the development work is done, the title may or may not be ready for use by the intended audience. If the title is for presentations (such as a sales representative to a client, or a corporate officer at a conference), it may simply be stored on the hard drive of a presentation system (such as a laptop) or even fit on a floppy disk. In these cases, the focus would be on controlling the environment: Where will the presentation take place? Will everyone be able to view it? What contingency plans are needed in case the equipment or application does not work? Can the lighting be controlled? Should audience handouts be provided? Even in the case of in-house presentations and training, it might be appropriate to use CD-ROM as the storage medium. Certainly, for mass market applications CDs are the most cost-effective medium.

In this and the next chapter, you learn about the production and distribution of multimedia titles. This chapter focuses on the production of CD-ROMs, and chapter 10 concentrates on CD-ROM, kiosk, and online delivery of multimedia titles.

Compact Disc

The ***compact disc (CD)*** provided the technological advance that revolutionized the multimedia industry in a way similar to the effect the printing press had on publishing. The CD allowed developers to provide interactive, feature-rich titles relatively easily. Compact discs offer many advantages:

- **Large storage capacity** CDs can hold more than 650 megabytes (MB) of data and 74 minutes of audio.

- **Durability** CDs are made of rigid plastic that is more sturdy than that used for floppy disks and tapes.

- **More difficult to copy** Unlike floppy disks, which are easily copied, CDs cannot be reproduced without expensive equipment.

- **Ease of distribution** Because of the light weight, size, durability, and non-volatility of CDs, they are easily distributed individually or in bulk.

- **Non-volatility** Unlike magnetic storage media (tapes and floppy and hard disks), CDs utilize laser technology that physically alters the surface of the disc. This makes accidental erasure of information difficult and therefore makes CDs attractive for archiving important data.

- **Low cost** Compact discs are relatively inexpensive to produce—$0.02 per gigabyte (GB)—compared with other media, such as floppy disks ($250/GB), hard drives ($150/GB), and removable storage cartridges ($35–$100/GB).

There are of course drawbacks to CDs. First, the access time is very slow compared with hard drives—145 milliseconds (ms) compared with 12 ms. This is especially important in multimedia titles, which often have large video, graphic, sound, and animation files to access and display. Second, many older computers do not have CD drives. This is becoming less of a problem as newer computers are shipping with CD drives installed. Third, CDs are read-only; that is, unlike hard drives, floppy drives, and various other magnetic-based media, the user cannot write to the CD. (The term *CD-ROM* stands for compact disc–read-only memory.) Some multimedia titles allow the user to type in and save information such as a name or an answer to a question; these titles are programmed to save the information to a hard drive.

The evolution of the current CD-ROMs has resulted from a collaboration of companies interested in establishing industry standards. Standards are important because they help the industry grow. If different CD manufacturers decided to make their own proprietary discs that worked only with certain drives, multimedia developers would have to make a decision as to which set of specifications to use. Business, education, and home users would have to decide which CD drive to buy, realizing that not all CDs would work with any given drive. Although significant standardization has occurred, there are different CD formats that have been developed throughout the years.

CD STRUCTURE

In order to understand the differences in CD formats, you need to be familiar with the basics of a compact disc. A CD is 4¾ inches in diameter, about 1 millimeter (mm) thick, and can hold approximately 650 MB of data. It consists of plastic coated with a thin layer of aluminum, which is in turn covered with lacquer. During the production, "pits" are created on the disc using a molding process; between the pits are areas called "lands." Together the ***pits and lands*** represent the digital coded data on the disc (see figure 9.1). When a laser beam is passed over the disc, the light is reflected by the lands and not by the pits. The reflected light is read by a sensor, and a signal is sent to the computer, which translates the signal into binary code (see figure 9.2). The pits and lands form a spiral track more than 3 miles long that starts at the inside (hub) of the disc and continues to the outside edge (see figure 9.3).

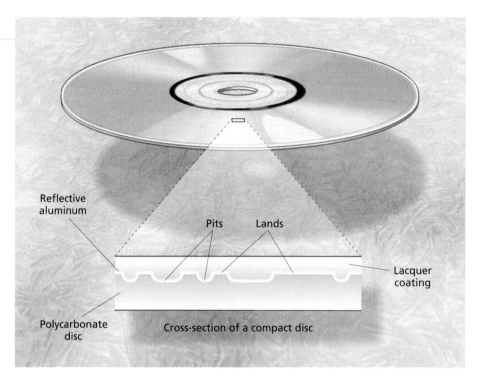

Figure 9.1

The pits and lands on a CD

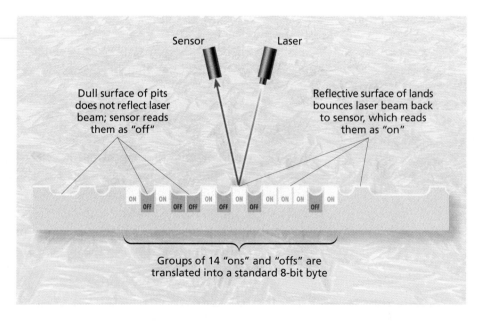

Figure 9.2

Reading a CD

Figure 9.3

The spiral track of a CD

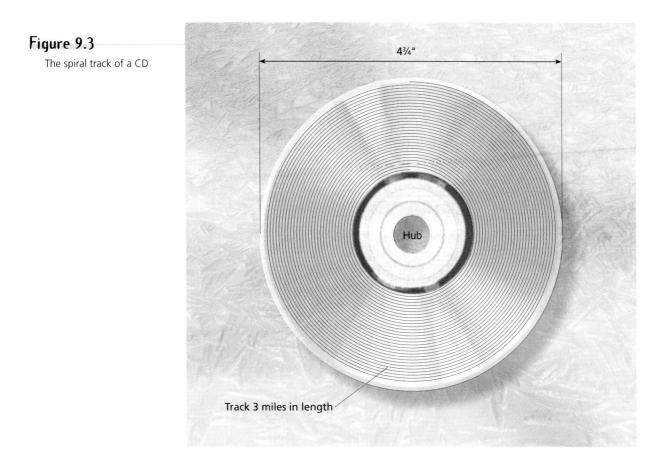

CD FORMATS

Following is a brief overview of several CD formats. A few of these formats include a reference to a color (red, yellow, or green); this refers to the color of the binding of the document in which the specifications for that format were first published.

CD Audio The first widely used compact discs were music CDs that appeared in the early 1980s. ***CD Audio***, or CD-DA, stands for compact disc–digital audio, the format for storing recorded music in digital form, as on CDs that are commonly found in music stores. The ***Red Book*** standards, created by Sony and Phillips, indicate specifications such as the size of the pits and lands, how the audio is organized and where it is located on the CD, as well as how errors are corrected.

Using compression techniques CD Audio discs can hold up to 75 minutes of sound. To provide the highest quality, the music is sampled at 44.1 kHz, 16-bit stereo. Because of the high-quality sound of audio CDs, they quickly became very popular. Other CD formats evolved from the Red Book standards.

CD-ROM Although the Red Book standards were excellent for audio, they were useless for data, text, graphics, and video. The *Yellow Book* standards built upon the Red Book, adding specifications for a track to accommodate data, thus establishing a format for storing data, including video and audio, in digital form on a compact disc. *CD-ROM* (compact disc–read-only memory) also provided a better error-checking scheme, which is important for data. One drawback of the Yellow Book standards is that they allowed various manufacturers to determine their own method of organizing and accessing data. This led to incompatibilities across computer platforms.

ISO 9660 In 1985 representatives from several influential companies in the technology industry, including Microsoft, Apple, and Sony, met at the High Sierra Hotel in Reno to establish standards that would allow CDs to be played across various computer platforms. They were looking to establish specifications for a common storage and playback medium that would be compatible with Macintosh, Windows-based, and UNIX-based computers. The results of this meeting became known as the High Sierra standards, which shortly thereafter were modified slightly and adopted by the International Standards Organization as *ISO 9660*. They included standards for the maximum number of levels in a directory structure (eight), and filename specifications (the DOS eight-dot-three convention).

CD-I Developed by Phillips in 1986, the specifications for *CD-I*, which stands for compact disc–interactive, were published in the *Green Book*. CD-I is a platform-specific format; it requires a CD-I player, with a proprietary operating system, attached to a television set. Because of the need for specific CD-I hardware, this format has had only marginal success in the consumer market. One of the benefits of CD-I is its ability to synchronize sound and pictures on a single track of the disc.

CD-ROM XA CD-ROM extended architecture, or *CD-ROM XA*, was developed by Microsoft, Phillips, and Sony in 1988. This extension of the CD-ROM format allows for interleaving data to enhance the playback of sound and video. It provides the same benefit of better synchronized sound and pictures as CD-I, but it works with the computer rather than dedicated equipment attached to a TV.

Photo CD The *Photo CD* was developed in 1992 by Kodak as a means of storing and viewing photos, slides, and film transparencies. The process involves scanning the photos with a high-end scanner, compressing the images, and writing them to a compact disc. More than 100 photos can be stored on a CD. The images are indexed, and thumbnails of several images can be displayed at one time. The most common use of the Photo CD is for archiving images and incorporating them into CD titles.

Kodak created its own player that connects to a TV for viewing the Photo CD images. The Photo CD can be played with a CD-ROM XA drive and a CD-I system. One significant feature of the Photo CD is that it can be written to more than once: That is, one roll of 35mm film could be processed and the images placed on the Photo CD; then, at a later date, another roll of film images could be added to the same CD. This is known as a ***multisession disc***. During the process of creating a multisession disc, a particular directory structure is created that provides access to each session's data. Not all CD drives can work with multisession discs. On those that can't, only the first session of images can be accessed. Photo CDs cost less than 50 cents per image and can be obtained from regular photo-finishing outlets.

CD-R Short for compact disc–recordable, the ***CD-R*** file format allows single CDs to be produced using a desktop CD-ROM recorder. The CD-R format is also referred to as compact disc–write once, or CD-WO. Compact disc recorders are laser-based systems that create one CD at a time—referred to as a ***one-off***. Depending on the recorder, CD-Rs can create discs for all the major CD formats discussed previously, including multisession discs that allow data to be written more than one time. Originally, these systems cost tens of thousands of dollars, but they are now available for less than $1,000, with blank discs costing under $10 each—less than 2 cents per megabyte.

The previous CD formats were developed specifically for a large market: business, education, and/or consumer. There are instances in which one or two CDs are all that is needed. The following are some everyday examples of practical uses of the CD-R format:

- Archiving printed material by scanning it into the computer and writing it to a CD
- Providing in-house multimedia training titles
- Distributing information to a small, select audience
- Creating a prototype of a CD title for testing
- Creating a premastered image file for delivery to a company for mastering and replication

A multimedia developer must decide which CD format(s) to use for any given title. The decision is based primarily on what CD drive and playback system(s) is used by the target audience.

The Production Process for CD-ROMs

In this section you learn about the process involved in creating CD-ROMs for mass distribution. It is assumed that the development process is complete and that the files for the multimedia title are located on a defragmented or newly formatted hard drive. The production process involves premastering, mastering, replication, labeling, and packaging.

PREMASTERING

The purpose of *premastering* is to create an exact image of what will be placed on the CD in an ISO 9660 or other CD format. The *image file* will include the data, filenames and directories, error detection and correction routines, indexes, and programs (such as an install program) that become part of the finished CD. Figure 9.4 shows a premastering program being used to create an image file.

Figure 9.4

A premastering program

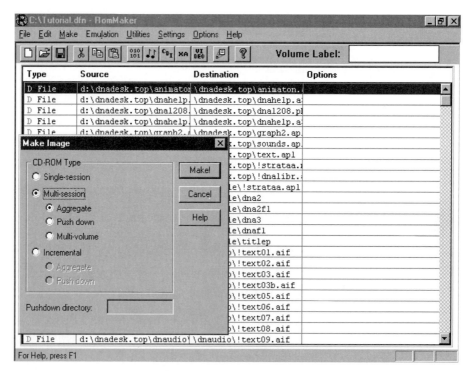

Premastering involves the following steps:

- **Organizing and naming the files** The files must be organized and named in such a way as to be compliant with the desired CD format. For example, ISO 9660 specifications allow only eight levels of directories, and the filenames must conform to the MS-DOS naming conventions.

- **Optimizing the file structure** This is necessary to speed up the retrieval time. Related files should be placed physically together on the disc. Those files that will be used often should be located near the beginning of the disc (near the hub), where access time is the shortest. Those files that will be used infrequently, such as install programs, should be located near the outer edge of the disc.

- **Simulating a CD using a hard drive** Retrieval time for a hard drive is faster than for a CD, and testing a multimedia title using a hard drive might not uncover performance shortcomings of the CD. There are premastering programs that will simulate the CD drive's seek and access time on a hard drive, allowing you to test the file structure for optimization and to determine problems that need to be corrected. Another testing process would be to use a CD recorder to produce a one-off for testing.

After successful testing, the premastered image file is delivered to a production company for mastering and replication. Various media can be used, including an external hard drive, external storage device (Syquest, Zip, DAT), or a CD-R. Although developers often do the premastering of a CD, companies that specialize in disc duplication can help with this step.

MASTERING AND REPLICATION

The next step is to create a master disc—a process known as ***mastering***. Because of the cost of the equipment, this is usually done by companies specializing in disc replication. Using a laser, the information on the premastered image file is burned onto a glass disc that has been covered with a light-sensitive material (see figure 9.5). The material is washed away, leaving pits and lands. The glass master is then used as a mold to create a stamper, which in turn is used to create the actual CDs. Once the stampers have been developed, the ***replication*** phase begins, and hundreds of discs per hour can be created.

Figure 9.5
Mastering a CD

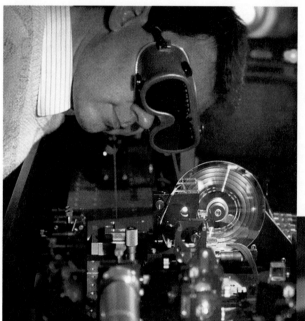

Left: A technician checks the position of a glass master about to be written by a laser. The laser is switched on and off at incredible speed to burn the pits that constitute data on the final CD. A printing master, called a stamper, is generated by making a reverse duplicate of the glass master, but in metal.

Below: The metal stamper is placed into an injection molding machine, where molten polycarbonate is compressed under tons of pressure, molding the plastic into a copy of the original master. In this photo, the new disc is still clear; in the next manufacturing step, the pitted side of the disc is coated with a thin layer of aluminum in a process called metalization.

Stamper

Newly pressed CD

Above: The CD, now with its familiar reflective surface, is covered with a layer of clear lacquer. This protective coating is cured, or hardened, by ultraviolet light.

Left: To ensure that the finished CDs will play back flawlessly, several quality assurance tests are performed during the manufacturing process.

The CDs are coated with a thin layer of aluminum so that the lands will reflect light. Finally, a coat of plastic and a layer of lacquer are applied to protect the disc.

LABELING AND PACKAGING

Labeling the CD is done through a silk-screening process. The label usually contains the name of the title, the name and logo of the company that developed the title, copyright information, and graphics. Graphics are an important part of the CD because they can add to its marketing appeal (see figure 9.6).

Packaging allows for bundling several items together, as well as aids in the marketing of a product. Some CDs need to be packaged in a large box because they require reference manuals or other documentation. Other CDs are packaged in a large box to help in merchandising the title. Large packages give the perception of added value, are useful in attracting attention on the retail shelf, and can provide sales information. Still other CDs are distributed with a minimum of packaging to reduce their costs and enhance their appeal to retailers that have limited shelf space. Figure 9.7 shows several packaging alternatives for CDs. The clear plastic **jewel box** has become a standard because of its relatively low cost, superior protection, and marketing features. A shrink-wrapped jewel box can show off the enclosed disc or promotional

Figure 9.6

Using the label to enhance the marketing appeal of a CD

Chapter 9   Producing Multimedia Titles   **193**

Figure 9.7
CD packaging alternatives

material and include printed text, such as instructions on running a setup program and warranty information. This material can be used for promotional purposes by utilizing graphics to attract potential customers (see figure 9.8). It is also useful in providing important information such as the computer system specifications required to run the CD.

Figure 9.8
The jewel box and printed insert

There is a trade-off in using a large package (called an "air package" because so much of the content is empty space) to hold a single CD. On the one hand, a large package can be extremely useful in differentiating one product from another, attracting attention, and thus increasing sales. Nearly half of the CD purchasing decisions are made at the point the customer is reading the package. This has led publishers to create intriguing packages in the shape of spaceships and trapezoids. On the other hand, this becomes a concern for the retailer whose shelf space is limited. Some publishers have tried smaller boxes, only to see their sales decline, and then returned to the larger ones. Other concerns with large packages are costs (up to $5 per package or more than five times the cost of the CD inside the package); and environmental effects of using paper products and excess packaging that will just be thrown away. The music CD industry forgoes air packages and uses jewel boxes and even paper envelopes almost exclusively. CD-ROMs may eventually follow this trend.

PRODUCTION COSTS

This section provides information on the costs of premastering and replication, as well as packaging and other services. Figure 9.9 shows costs for premastering and replication of a limited number of discs; notice that there is a difference in cost based on which medium is used to deliver the files. For example, if the multimedia developer delivers the files on a hard drive, the cost for premastering a single disc is only $99. If the files are delivered on multiple tapes, however, the cost is $159. It is possible to have multiple discs created using the premastering process and thereby eliminate the mastering and replication costs. This would be cost-effective only for a small number of CDs, however. Figure 9.10 shows costs for mastering and replicating large quantities of discs; notice that there are quantity discounts provided for replication as well as premiums for quick turnaround times. That is, if the turnaround time for mastering is five days, the cost is as low as $250, whereas if the mastering needs to be done in one day, the cost can be as high as $1,800.

Companies that specialize in CD-ROM mastering and replication can provide additional services, such as layout and design of printed materials that would be included with the CD; drop-shipping titles to a warehouse; and archiving of stampers. Working with a full-service CD-ROM production company can mean that once the premastered image file and label designs are delivered, the developer need not handle the finished product.

Figure 9.9

Costs for premastering and replication of small quantities of discs

Item	Cost Each
Single cartridge/hard drive/disc to CD-ROM	$99
Backup tape to CD-ROM	$129
Multiple cartridges/tapes to CD-ROM	$159
Additional copies	
1	$75
2–4	$60
5–10	$45
11–19	$30
20+	$25
Data conversion	$75/hour

Figure 9.10

Costs for mastering and replication of large quantities of discs

Replication				Mastering Turnaround Time		
Run Size	Disc	Jewel Box	Shrink Wrap	10 Days	5 Days	1 Day
100	$1.95	$ .29	$ .05	$700	$850	$1,800
500	$1.50	$ .29	$ .05	$450	$700	$1,800
1,000	$1.02	$ .29	$ .05	$250	$600	$1,800
10,000	$ .90	$ .28	$ .04	N/C	$250	$1,400

To extend what you've learned, log on to the Internet at
http://www.thomson.com/wadsworth/shuman
You will find a wide variety of resources and activities related to this chapter.

key terms

CD Audio	mastering
CD-I	multisession disc
CD-R	one-off
CD-ROM	packaging
CD-ROM XA	Photo CD
compact disc (CD)	pits and lands
Green Book	premastering
image file	Red Book
ISO 9660	replication
jewel box	Yellow Book
labeling	

review questions

1. **T F** An advantage of CDs is the relatively large storage capacity.

2. Together the _____ and _____ represent the digital coded data on a CD.

3. **T F** A drawback of CDs is their slow access speed relative to hard drives.

4. CDs that comply with the _____ (also know as High Sierra) standards can be played using various computer platforms, including Macintosh, Windows-based, and UNIX-based.

5. The _____ was developed by Kodak in 1992 as a means of storing and viewing photos and slides.

6. The purpose of _____ is to create an exact image that will be placed on a CD.

7. **T F** Nearly half of the CD purchasing decisions are made at the point the customer is reading the package.

8. **T F** Companies that specialize in CD-ROM mastering and replication seldom provide layout and design of printed materials that would be included with the CD.

9. **T F** CD-I stands for compact disc–Internet compatible.

10. **T F** Compact discs are most cost-effective to produce in large quantities.

projects

1. Contact a company that masters CD-ROMs and prepare a report with the name and address of the company as well as the following details:

   - Services provided
   - Costs involved for various services
   - Timelines required for mastering and replication
   - Clients served
   - Examples of CD titles they have produced
   - Titles they are most proud of and why
   - What is the most difficult part of the client/provider relationship
   - Trends they see in the industry

   Prepare a brief oral presentation of your report and be ready to present it to your class.

2. Study several multimedia CD-ROM titles and prepare a report that includes each title's name and price as well as the following details:

   For the packaging:
   - Type of packaging
   - The Information provided on the package
   - How the package contributes to the promotion of the title
   - How the package could be improved

   For the CD label:
   - The information provided on the label
   - How effective you feel the label is in marketing the title—explain
   - How the label could be improved

   For the jewel box insert:
   - The information provided on the insert
   - How the insert is used to promote the title
   - How the insert could be improved

   Prepare a brief oral presentation of your report and be ready to present it to your class.

Distributing Multimedia Titles

AFTER COMPLETING THIS CHAPTER YOU WILL BE ABLE TO:

- Describe the marketing issues related to publishing CD-ROM titles
- List and describe the major distribution models
- Describe how multimedia titles are distributed online
- Discuss the issues related to delivering multimedia using kiosks

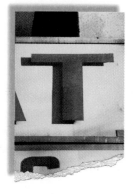

HERE are nearly 5,000 CD-ROM titles and more than 2,000 developers, and both numbers are growing. The average cost of producing a commercial-quality title is nearly $400,000, which means that a single title would need to sell 20,000 copies just to break even. It is estimated, however, that eight out of ten CD-ROM titles lose money. No title, even an award-winning one, is assured of commercial success; but a multimedia title that is not properly "delivered" to the intended user is doomed to failure.

During the planning process for a multimedia title, the ***target audience*** is identified. Identifying the intended audience is crucial to the creation of the title, helping to determine its elements, content, design, and functionality. Identifying the intended audience is also important in determining how the title will be delivered to the user. For in-house training or presentation titles, the delivery is fairly straightforward, ranging from a conference room or classroom to installing the title on desktop computers or a company network. Delivering commercial titles to the user involves an entire marketing process. This chapter focuses on how companies market commercial multimedia titles, including issues of pricing, promotion, and distribution, and the factors involved in selecting the medium (CD-ROM, online, kiosk) to deliver the multimedia title. In this chapter a distinction is made between those involved in creating a multimedia title—developers—and those involved in distributing titles—publishers. ***Developers*** create the title and prepare it for mastering and replication. ***Publishers*** are involved in the marketing and particularly the distribution of multimedia titles. In many cases, the same company both develops and publishes the title.

Distributing Multimedia Titles on CD-ROM

MARKETING CONSUMER TITLES

Multimedia publishers face several market-related hurdles that must be overcome. First, the industry is new and most consumers have little or no experience with multimedia titles. Consumers are reluctant to spend money on something they are unfamiliar with and, unlike video, book, and music titles, there are few sources for reviews, previews, or ratings of multimedia titles. In addition, most retailers are not set up to demonstrate CD-ROM titles. Second, there has been a high rate of returns (up to 25 percent) of CD-ROMs by first-time buyers who purchased a title and then were unable to run it on their computers. This is aggravated by the fact that retail employees may not have the training to assist customers with technical problems. Third, there is a lack of good content. Many titles are simply books and movies that have been repurposed without much imagination and without taking advantage of the benefits of multimedia. Fourth, and perhaps most crucial, there are too many CD-ROM titles vying for too little **shelf space**. Indeed the single most important concern when distributing consumer titles is retail display area—and the lack thereof. To overcome these hurdles, a publisher must develop a marketing plan that involves coordinated product, promotion, pricing, and distribution strategies.

PRODUCT STRATEGY

Multimedia development companies realize that creating only one product, even if it is commercially successful, will not sustain a company in the long run. Companies develop mission statements that often include goals of becoming leaders in a particular field (such as children's titles) or niche market. In a highly competitive field, a company's multimedia titles must be differentiated from the others. One way to differentiate would be to create a character such as Arthur in the series of titles put out by Living Books (see figure 10.1). Each of the Arthur titles has the same "look and feel," objectives,

Figure 10.1

Using a character to differentiate a product

main character, and theme. What changes is the setting and the type of adventures Arthur experiences. Another way to differentiate a product is by creating high-quality graphical images, rich content, or expensive elements such as 3-D animation and video.

Another product strategy is to develop brand-name recognition by the target market. Names such as Microsoft or Brøderbund are immediately recognizable and generate confidence in the consumer. Start-up companies, of course, do not have such name recognition, but by adhering to certain development standards and specifications and by entering into a licensing agreement with Microsoft for example, they could carry the Microsoft logo on their product packaging and thus generate confidence in the consumer by association.

Whatever product strategy is used influences and is influenced by the publisher's strategies for promotion (Should advertising be focused on brand recognition or individual titles?), pricing (higher prices usually connote quality), and distribution (Should the titles be sold directly to the consumer or through specialty retailers?).

PROMOTION STRATEGY

Retailers are more inclined to stock a title that has an aggressive promotional plan that includes an advertising campaign, point-of-purchase material, and publicity. They are interested in advertising that will stimulate demand for a product and pull the customer into their store. The question is: Who pays for the advertising? One technique is for the retailer and the publisher to share the cost through a ***cooperative advertising*** program (see figure 10.2). This involves promoting the multimedia title and the retailer in the same advertisement in order to generate demand for the title and traffic for the store. Even this type of shared-cost advertising is too expensive for all but the larger publishers.

Point-of-purchase materials such as aisle displays can be tied in with an advertising campaign and are useful in encouraging impulse buying (see figure 10.3). Aisle displays can increase the sales of a product by 25 percent or more, even if the price is not lowered. The product receives more exposure, and the consumer perceives some added value for a product that is on display. Point-of-purchase displays must be designed, manufactured, and shipped to retailers, however, and there may be no guarantee that the displays will be set up and maintained by the retailer once they are received.

Publicity comes from stories that appear in various news media. For multimedia titles this is usually magazines, newspapers, and trade publications. The stories often appear as reviews of new products (see

Chapter 10 Distributing Multimedia Titles

Figure 10.2
An example of a cooperative advertisement

Figure 10.3
Point-of-purchase display

Figure 10.4
Reviews of new CD-ROM titles

CD-ROM

Lost in Space

BY RON WHITE

The best thing about science fiction movies are the previews—giant ants pinching a soldier in half, King Kong biting off someone's head, death rays blasting from Martian spaceships. In contrast, the movies themselves are always disappointing—you get a lot of padding and dumb dialog to kill the time until the special effects kick in. But I keep renting the videos. I also keep coming back to science fiction CDs. Unfortunately, a lot of sci-fi CDs are also padded, with dull mouse-groping and multimedia that substitutes for the real meat of science fiction: imagination. Here's a peek at what sci-fi fans can look forward to or avoid on CD.

RAY BRADBURY'S MARTIAN CHRONICLES ○○○○
Scientists now tell us there once may have been life on Mars. Ray Bradbury told us that decades ago. His classic sci-fi book, *The Martian Chronicles*, is a wonderful combination of imagination, storytelling, and poetry—a lot of which surprisingly survives the translation to CD-ROM. Like most adventure games, this one has a lot of riddle solving and aimless poking around as you try to uncover the extinct Martians' secret for peace. Bradbury's words and characters crop up from time to time in the form of Martian telepathic images that have outlasted the Martians themselves. In true science-fiction style, not everything is what it seems to be. The game is more awkward to maneuver through than it should be, which means you spend a bit more time clicking your mouse than getting into the adventure.

Ray Bradbury's Martian Chronicles Adventure Game / Byron Preiss Multimedia / (800) 910-0099, (212) 698-7000 / $60 est. street price / *Reader Service No. 719*

Life on Mars: Discover the classic *Martian Chronicles* on CD-ROM.

 TIE FIGHTER COLLECTORS ○○○○
Here's a chance for your dark side to get in the pilot's seat. This sequel to X-Wing has you piloting a Tie space fighter in the service of the evil Empire from *Star Wars*. Of course, when it comes to flying simulation battles, what counts is zapping the other guys before they zap you. (You can worry about your conscience once you make it back alive.) Tie Fighter's 3-D simulation gives you all the action you can handle, once you get to it. There's an awful lot of role-playing to go through before you get your hands on the joystick, and you have to learn a couple dozen keystrokes to handle your fighter. I prefer something like Microsoft Fury, in which you're dropped into the middle of the action and have to fly by the seat of your pants. (And the Windows 95 version of Fury is easier to set up than the DOS-based Tie Fighter.) Tie Fighter is for experienced sim fighters out of my league who want all the challenge they can get.

Use the Force: In Tie Fighter you're the bad guy, but who cares?

Tie Fighter Collectors / LucasArts Entertainment / (800) 985-8227, (415) 472-3400 / $30 est. street price / *Reader Service No. 720*

 I HAVE NO MOUTH AND I MUST SCREAM ○○○○
The bad boy of science fiction, Harlan Ellison is a snarly iconoclast whose credits include *Star Trek* and some of the darkest stories you'll find in sci-fi. His short story "I Have No Mouth and I Must Scream" is as cheerless as a bus station at two in the morning. The DOS

○○○○○ *You're beamed up to the Mother Ship.* ○○○○ *You discover a new life form.*
○○○ *A life form discovers you.* ○○ *It's William Shatner.* ○ *Zippers in the monster suits.*

figure 10.4). Publicity, assuming it is favorable, can be extremely valuable to a publisher for several reasons. First, it is unbiased and therefore lends credibility to the reviewer's conclusions. Second, it can be leveraged by including reviews in advertising, on point-of-purchase displays, and on packaging. Third, it can be directed toward the desired audience. For example, a company that has created a multimedia title called The Castles of Scotland could send review copies to several travel magazines. Fourth, publicity is free. In order to generate publicity, multimedia publishers prepare and distribute to selected reviewers **press kits** containing product information and a full version (not a demonstration version) of the title.

PRICING STRATEGY

With more and more multimedia titles being produced, prices are becoming extremely competitive. Prices for CD-ROM titles for the home market have steadily declined from the $70-to-$90 range to around $20 to $50. Pricing is heavily influenced by cost and competition. Traditionally, companies have determined a *suggested retail price (SRP)* and provided discounts to different buyers. Wholesalers might receive a 50 percent discount and retailers a 35 percent discount off of the suggested retail price, which is usually above the *street price*—the price actually paid by the consumer. This puts pressure on retailers to "turn over" their inventory almost monthly to accommodate the lower margins. This is one reason why retailers stock only the best-selling titles.

In many cases, publishers have not approached the pricing of their products in a way that will cover their entire costs. Some have included only development and production costs and not provided for marketing costs or company overhead. The following example shows how many copies of a title need to be sold to cover the costs of development and production on a CD-ROM title.

Break-even example:

Development costs: $400,000

Per-unit production (CD-ROM, packaging, labeling, shipping): $2

Retail price: $40

Discount provided to wholesaler: 50 percent ($20)

Publisher's per-unit revenue: $20 ($40 – $20)

Per-unit amount available to cover development costs: $18 ($20 – $2)

Number of units needed to sell to cover total development costs: 22,222 ($400,00 ÷ $18)

In this example more than 22,000 units would need to be sold to break even—and this includes neither promotional expenses nor company overhead. For every $100,000 in promotional and overhead expenses, another 5,555 units would need to be sold; and this would not include any profits for the company.

Initially, a publisher may decide to price a title to cover only development and production costs as a way to keep the price competitive and obtain a share of the market. Eventually, the title must be priced to cover all costs and provide for a profit. In the previous example, one way to keep the price low and return more to the publisher would be to eliminate the wholesaler and go directly to the retailer or the user. Distribution strategy, then, becomes a major influence on pricing.

DISTRIBUTION STRATEGY

Multimedia is a new industry that affects many fields: entertainment, video, music, computer, and education, to name but a few; and it involves many players: book publishers, film producers, software companies, small multimedia developers, and so forth. Given the eclectic nature of the industry, it is not surprising that several distribution models are being tried—with varying degrees of success. In theory a CD-ROM publisher has several options when determining how to distribute consumer multimedia titles. In practice the options are restricted by the resources of the publisher, especially in terms of promotional support. The ultimate test of a title is its demand, and sales are heavily influenced by retail exposure, advertising, and point-of-purchase displays.

Figure 10.5 shows three traditional distribution models that have been used for delivering CD-ROM titles to the consumer market. The decision facing the CD-ROM publisher is which channel or channels to use. Traditionally, companies that make consumer goods that are purchased in a retail store sell their products to a wholesaler, who in turn sells to retailers. In some cases, the company sells directly to retailers. Whatever model is used, there needs to be a *value-added* effect as the product moves through the distribution channel; that is, every link in the distribution chain needs to make some contribution to the product and its perceived value to the consumer. Following are the pros and cons of the various distribution models.

Wholesalers *Wholesalers* buy from multimedia publishers and resell to others, including retailers, catalog companies, and corporate and education markets. Wholesalers can provide services such as controlling inventory, processing returns, and giving technical support. A primary benefit for multimedia publishers is that wholesalers offer access to a worldwide network of retailers. In fact, this may be the only way in which a publisher can get access to retailers. There are a few large wholesalers that specialize in computer-related products, including hardware, software, and accessories. These companies, such as Ingram Micro, Inc., which is the largest software distributor, also handle CD-ROM titles. Other wholesalers have either created multimedia departments or specialize in particular multimedia titles. Baker

Figure 10.5

Three traditional distribution models

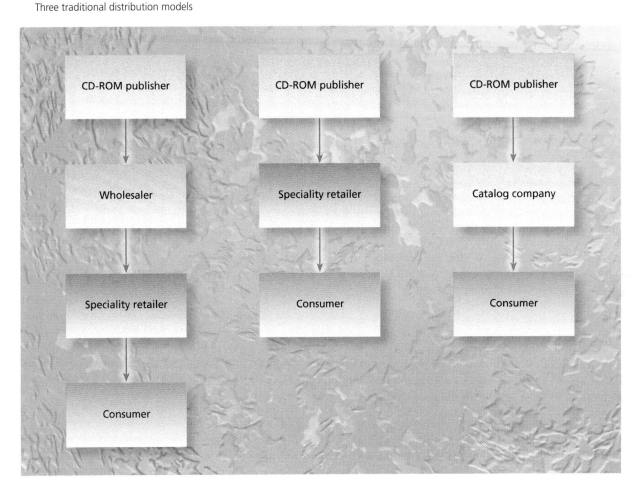

and Taylor, one of the largest book, video, and software distributors, reorganized to create an entertainment division, trained its salespeople to handle multimedia products, and entered into a licensing agreement with Parent's Choice Foundation to provide customers with unbiased advice on children's multimedia titles. The Davidson company is a wholesale distributor specializing in children's titles, and CMEA is a wholesaler specializing in medical titles.

There are drawbacks to using wholesalers. Because wholesalers carry thousands of products, there is little likelihood that they will actively promote any one CD-ROM title or work to obtain retail shelf space. Rather, they may just list it in their catalog along with related products. Also, the publisher would have to substantially discount the suggested retail price to provide an incentive for the wholesaler to handle the product. Discounts of 50 percent or more are common for wholesalers.

Retailers One of the best ways to reach the consumer market is through *retailers*, although a typical retail outlet may stock only 100 to 200 titles out of the thousands that are published each year. A retailer selling CD-ROM titles is interested in maximizing sales per square foot of floor space by increasing inventory turnover. Thus, a primary criterion for stocking a particular title is ***demand***.

Forecasting demand for a new title is risky, and retailers look at a company's track record and promotional plan to try to gauge the consumer appeal of a new product. Another important consideration for a retailer in determining which titles to stock is pricing. The publisher's pricing structure has to allow a margin of 35 to 50 percent for the retailer. Even higher margins may not generate substantial revenue, because of the extreme price competition that retailers face. Retailers are also hesitant to do business with a company that has only one or two products, rather than a complete product line. They want to limit the number of companies they work with, because every supplier that a retailer uses costs money in terms of contacts, invoicing, shipping, and time. Therefore a publisher with a limited number of titles may have to work through a wholesaler who can combine titles to make a comprehensive product offering to the retailer.

Because of the tremendous demand for shelf space (only one in 15 titles finds its way to a retailer's shelf), retailers often obtain market development funds, called slotting fees, from publishers to stock a title. In other words, the publisher pays to have the product carried by the retailer.

Superstores and Specialty Retailers Originally, CD-ROM titles were sold through computer stores such as Egghead, CompUSA, and Computer City. This was because those who had computers shopped at these outlets for software. Eventually, mass merchandisers such as Wal-Mart, K-Mart, and Target stores began carrying CD-ROM titles. Now book, toy, video, and music stores as well as other specialty retailers are selling multimedia titles. More than 25 percent of independent bookstores carry CD-ROM titles. They average about 50 titles per store and focus on reference and children's titles. Most national bookstore chains such as Barnes & Noble and Borders stock reference titles that they feel fit with their other product lines and with their customer base. Bookstore chains devote more of their space to multimedia titles than independent stores (6 percent and 2 percent, respectively). Toys R Us signed an exclusive contract to have distributor New Media Express provide nearly 150 multimedia children's titles to their 600 plus stores. Children's titles are also showing up in early-learning superstores such as Zany Brainy, which has more than 30 outlets. Hardware, sporting goods, and grocery stores are stocking CD-ROMs relevant to their product lines and clientele. Blockbuster Video carries games, and Blockbuster Music carries interactive music titles. Specialty retailers offer a good distribution alternative for multimedia publishers, especially when the retailer's customer base matches the publisher's target market.

DISTRIBUTION ALTERNATIVES

Given the highly competitive nature of the consumer multimedia market and the difficulty of obtaining retail shelf space, other distribution models are being used. These include direct mail, bundling, catalog sales, rentals, and partnerships.

Direct mail Publishers may find that mail-order sales is a cost-effective way to enter the market, especially if mailing lists are available for their target audience. For example, a company that creates an interactive golf title could purchase the names and addresses of subscribers to *Golf Digest* magazine and send a direct-mail piece to them. Mailing lists, as well as printing and mailing costs, are relatively inexpensive when compared with other advertising media such as newspapers and magazines. Purchases by 5 percent of those receiving a mailing are typical. The important thing is that the mailing list matches the target audience.

Bundling *Bundling* involves the distribution of a multimedia title with some other product such as a new computer or upgrade kit. For example, several top-selling titles including Myst, Rebel Assault, Magic Carpet, and SimCity 2000 are bundled with the Diamond Multimedia Kit; and various titles including Compton's Encyclopedia, Sports Illustrated for Kids, Jack Nicklaus Golf, and Rapid Assault are bundled with an IBM multimedia computer. This type of distribution is called **OEM** marketing, because the publisher works with the "original equipment manufacturer." The advantage for the publisher is that essentially no marketing costs are involved other than negotiating with the hardware manufacturer. This is a quick way to get a product out to a new market in hopes of establishing brand identification and follow-up sales of new versions. A disadvantage is that the large discounts required (as high as 70 to 80 percent of the selling price) leave very little return on each unit.

Catalog sales Several companies, such as Micro-warehouse, PC Connection, and Multiple Zones International (see figure 10.6), publish monthly catalogs of computer hardware, software, and CD-ROM titles and distribute these to selected markets. Multiple Zones, for example, publishes the *Mac Zone* (over 2 million circulation) and the *PC Zone* (over 1 million circulation) as well as corporate, education, home, and international catalogs. These companies can act as the direct-sales arm of a publisher by taking

Figure 10.6

A catalog listing of CD-ROM titles

phone and fax orders and shipping the product directly to the user from a warehouse. The companies "sell" advertising in their catalogs to multimedia publishers. Rates for a one-sixth-of-a-page ad start at approximately $1,000. This can be an extremely cost-effective way to reach a targeted market, but it is important to understand that this type of marketing needs to be done on a consistent basis. Continual advertising reinforces the promotional message, keeps the product's name in the consumer's mind, and gives the impression of a large, well-funded company.

Rentals Some retailers are experimenting with renting CD-ROMs. Blockbuster, a national chain of video and music stores, rents and sells CD-ROM game and entertainment titles. This not only helps the retailer diversify its product line, but also helps position it for the eventual expansion into movies on CDs. While Blockbuster may handle 200 titles, Booktronics in Houston carries several hundred titles for sale and rent. Booktronics also provides a demo room for customers to sample titles, and trains employees to handle technical questions. Publishers who are considering allowing their titles to be rented need to consider the effect on sales. Will a person who rents a title eventually buy it? Rentals are a way to "demo" a title, and users who find a title rich in content (reference works) or compelling enough to be played again and again (games) may eventually purchase the title.

Partnerships With the incredible growth of multimedia developers and the distribution problems facing them, it is no surprise that partnerships, joint ventures, and acquisitions are fairly common. Many of these bring together a developer and a distributor or content expert. The developer might benefit by gaining instant access to the distributor's market. This minimizes the developer's marketing costs and allows the developer to concentrate on creating titles. In turn, the developer gives up control over how the product is marketed and in some cases must allow the partner to influence the design and content of the title. Financial arrangements vary, from having a distributor fund the development costs to licensing a title and paying a royalty on each unit sold. Following are some examples:

- Random House New Media (distributor) and Brøderbund Software (developer) collaborated on the Living Books series including the Dr. Seuss titles

- Times Mirror (distributor) acquired Ehrich Multimedia (developer) which created Food & Wine Tasting

- Microsoft (developer) formed a joint venture with Scholastic (content provider) to produce a series of children's education titles including The Magic School Bus Explores the Solar System

- Meredith corporation (magazine publisher) and Multicom (multimedia company) collaborated on the Better Homes & Gardens children's title Dandy Dinosaur

- Mercury Magazine (distributor) is working with StarPress Multimedia (developer) to distribute the Sports Illustrated Multimedia Almanac

- Alpine books (distributor) has an arrangement with Media Mosaic (developer) to distribute a mountain-biking title

- Reader's Digest (content provider) has joined with InterMedia Active Software, Inc. (developer) to create a series of crossword puzzle CD-ROMs

MARKETING NON-CONSUMER TITLES

Non-consumer titles, such as reference works used by libraries, corporate training titles, and educational titles directed at the college market, make up the vast majority of CD-ROM sales. They are less complex to market than consumer titles, do not require shelf space, and are not as prone to seasonal sales. The target audience is usually smaller, more defined, and easier to reach. Thus the multimedia publisher can take a more direct approach to distribution. Large wholesalers such as Ingram Micro, who sell to retailers, also have corporate divisions. Other distributors, such as Software Services, specialize in sales to educational institutions or government agencies. Because of the fewer number of potential buyers, telemarketing and direct-mail campaigns can be effective. Trade shows that allow vendor demonstrations are also a useful way to reach specific industries.

Distributing Multimedia Titles Online

In April 1994 here were 1,000 sites on the Internet. Eighteen months later there were 110,000 sites. In 1996 there were more than 20 million people worldwide who had access to the Internet. One prediction is that this number will exceed 50 million by the end of the century. In a survey of Internet users, 82 percent said that online services will change the way they entertain themselves, and 91 percent said that such services will change the way they learn.

The Internet is quickly becoming a major delivery medium for multimedia. The World Wide Web provides all of the elements of multimedia, including sound, animation, video, and hyperlinking—and the content can be updated as often as desired. Thus the advantage of **online distribution** is the vast audience and the ability of the developer to constantly refresh the content to keep it timely. The drawback of delivering multimedia online is its slow speed. The fastest commonly used modems transfer data at a fraction of the speed of CD drives, so a major disadvantage of distributing online is the inability to deliver large files quickly. On the other hand, although the information on a CD-ROM can be accessed relatively quickly, once a CD is mastered no data can be added. This means that whatever is on the CD when the user obtains it will not change. For game titles this might not pose a problem, but reference titles could eventually become obsolete.

Thus the Internet and CD-ROMs are complementary—the strength of one overcomes the weaknesses of the other. For example, Microsoft's reference title Complete Baseball provides historical data and team and player statistics. Those who have registered copies of the CD-ROM can access daily game updates through the Microsoft Network for $1.25. This distribution process has significant value for the education and corporate-training markets. An online course could be developed and a CD-ROM sent to those who enroll. The CD-ROM would have course materials such as case studies, reference works, tests, tutorials, virtual labs, and other tools. The online component would allow instructors to give assignments involving material on the CD-ROM, establish "chat rooms" for student discussion, and update and augment materials as needed.

If a publisher includes online distribution in the marketing mix, a decision needs to be made on how to structure the online connection. The developer could install a server and establish a home page that could be accessed by users, but consideration would need to be given to how the system would be maintained and how often the content would be updated.

Kiosk-based Multimedia

As you recall from chapter 1, a kiosk is a computer-based system used to provide information and/or conduct transactions. The most widely used kiosks are bank ATMs (automated teller machines). Although ATMs are not used to deliver multimedia, kiosks in general are a good medium for multimedia because they can be configured to accommodate the needs of the user and the developer. For example, a videodisc or a large hard drive could be added to the system to accommodate video; a printer could be part of the kiosk to provide a hard copy of desired information; and if the content needed to be updated, the kiosk could be linked to a network. Kiosks are expensive. A stand-alone system that has a secure cabinet, high-end computer, large hard drive, fast video card, large touch-screen monitor, and sufficient RAM can cost $15,000 or more.

There are two types of kiosks, informational and transactional. ***Transactional kiosks*** provide for user input. Examples of transactional kiosks are ATMs, school registration systems, and airline self-ticketing systems. In most cases, the user must enter information (name, ID number, PIN) in order to complete a transaction. Because this often requires more then just pressing buttons on a touch screen, an input device such as a keyboard, numeric pad, or card reader may be necessary. Fortunately, a touch screen can display a keyboard, so no peripheral unit is needed. Examples of ***informational kiosks*** are those found in shopping malls, libraries, and museums, which allow the user to access information such as store location and hours; city services and employment opportunities; and art collections and artist biographies.

The success of a kiosk is to a large extent determined by its location: It needs to be accessible to the intended audience. In most cases, this means locating the kiosk in a high-traffic area that is convenient to the potential user. Shopping malls, college bookstores, bank buildings, airport terminals, libraries, and museum lobbies are examples of practical locations. The kiosk's location will often have an effect on its features. For example, sound might be considered a desirable element that could be used to attract people to an amusement-park kiosk, but if the kiosk were located in a library reading room, the use of sound would be inappropriate. Developers of a kiosk to be situated at the city zoo would have to consider children among the intended users; whereas those creating a night-life directory of clubs and restaurants for urban tourists would not. Other access issues, especially for kiosks in public places, are how the kiosk will accommodate those who are wheelchair bound, visually impaired, or non-English-speaking. Other considerations when developing a kiosk are how it will be maintained and how the content will be updated. Users who experience kiosks that malfunction or have outdated content will not return to use the system.

An example of the use of a kiosk to sell CD-ROMs is a program developed by Follett Campus Resources. Follett sets up kiosk systems in college bookstores that allow students to preview more than 50 CD-ROM titles sold by academic publishers such as Irwin, International Thomson, and Prentice-Hall. The students can search for a particular CD by publisher, name, discipline, or topic. The bookstore leases the kiosk from Follett for a fee that covers maintenance and monthly updates of the titles. Follett also receives a subscription fee and title fee from each publisher.

To extend what you've learned, log on to the Internet at

http://www.thomson.com/wadsworth/shuman

You will find a wide variety of resources and activities related to this chapter.

key terms

bundling	publisher
cooperative advertising	retailer
demand	shelf space
developer	street price
informational kiosk	suggested retail price (SRP)
OEM	target audience
online distribution	transactional kiosk
point-of-purchase	value-added
press kit	wholesaler
publicity	

Chapter 10 Distributing Multimedia Titles

review questions

1. **T F** About eight out of ten CD-ROM titles lose money.

2. **T F** Historically, CD-ROM titles have had a low (less than 5 percent) return rate to retail stores.

3. **T F** Aisle displays can increase the sales of a product by 25 percent or more.

4. Companies send review copies of their product to magazine editors in order to generate _____.

5. A retailer's primary criterion for deciding to stock a particular multimedia title is _____.

6. **T F** Throughout the years prices for CD-ROM titles for the home market have remained relatively stable.

7. _____ involves distribution of a multimedia title with some other product such as a new computer or an upgrade kit.

8. _____ can act as the direct-sales arm of a multimedia publisher by taking phone, fax, and online orders and shipping the product directly to the user from a warehouse.

9. **T F** The Internet is becoming a major delivery medium for multimedia.

10. The two types of kiosks are _____ and _____.

projects

1. Find three articles (magazine or newspaper) that review three different multimedia titles. Develop a report that includes the following specifics for each:
    - The publication and reviewer's names
    - The name and developer of the title
    - The type of title and intended audience
    - The platforms the title runs on
    - What was favorable in the review
    - What was unfavorable in the review
    - The overall rating or opinion of the reviewer
    - Whether the publication was appropriate for the intended buyer of the title
    - Other information that would have been useful to a potential buyer

    Prepare a brief oral presentation of your report and be ready to present it to your class.

2. Visit a retail store that sells CD-ROM titles. Develop a report that includes the following specifics:

   - The type of titles carried by the store
   - Whether the type of titles is appropriate for the store's customers—explain
   - The number of different multimedia publishers represented
   - The type of point-of-purchase displays, if any, for the titles
   - The most commonly used price points (for example, $19.95, $39.95, $69.95) and the correlation, if any, between the price points and the type of title

   Prepare a brief oral presentation of your report and be ready to present it to your class.

3. Select several multimedia titles that are distributed through retail stores as well as three of the following: specialty stores, magazines, online, and catalogs. Develop a report comparing the titles using the following criteria:

   - Price
   - Shipping and handling costs
   - Return policy
   - Availability

   Prepare a brief oral presentation of your report and be ready to present it to your class.

4. Search the World Wide Web for at least three companies selling CD-ROM titles online. Develop a report that includes the following specifics:

   - The Web site name and address (URL)
   - How you were able to locate the site
   - What the process is for searching for a title from the site
   - What the process is for ordering a title
   - The method(s) of payment used
   - Evaluation of the legitimacy of the site—explain
   - Price of the title online as compared to the retail price
   - Whether you would order a title from this site—explain

   Prepare a brief oral presentation of your report and be ready to present it to your class.

Part V

Multimedia Issues and the Future of Multimedia

The Internet and the World Wide Web

AFTER COMPLETING THIS CHAPTER YOU WILL BE ABLE TO:

- Describe the Internet and the World Wide Web
- Describe the factors that contributed to the growth of the Web
- Specify the considerations in developing multimedia titles for the Web
- Describe the common ways to develop multimedia titles for the Web

What Are the Internet and the World Wide Web?

The **Internet** is a vast communications system linking computers around the world. When two or more computers are linked together, it is called a **network**. Networks are often configured with a more powerful computer, called a **server**, that controls the network and provides a large storage capacity. The other computers on the network, called **clients**, allow users at remote locations to access the programs and data on the server. Businesses, government agencies, schools, and other organizations have been using networks for decades. Networks allow individuals within organizations to communicate and share information. *The Internet is a network of networks* (see figure 11.1). It was originally developed by the U.S. government and educational research institutions. The government, especially the Defense Department, was interested in having a communications system that would link together different types of computers and allow government-sponsored research to be shared. Universities were interested in having a communications system that would assist in collaboration on research projects and dissemination of research findings. Until the early 1990s, there was not a

Figure 11.1

The Internet—a network of networks

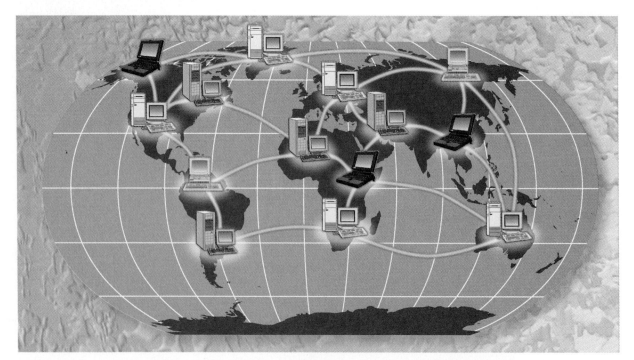

general interest in the Internet by companies or individuals. Two events, however, helped cause a surge in the use of the Internet.

First was the development of two computer features: a visual interface that, with a browser, allowed the user to easily search for information; and a formatting standard specifically designed for the Internet, called ***Hypertext Markup Language (HTML)***. Initially, the Internet was text-based: There were no photographs, sounds, animations, or video. In order to find data on the Internet, users had to navigate through a menu structure that often was several levels deep. Cryptic commands such as **ls** (to display a list of files) and **cd** (to change directories) were used. Figure 11.2 shows a menu structure used to access information on the Internet. In 1993 a program called Mosaic was developed and became the first popular visual interface that could be used to display Web pages. ***Web pages*** are documents that are written in HTML and form the basis for the ***World Wide Web (WWW)***. Figure 11.3 shows a Web page with a visual interface used to display information and navigate to other Web pages. HTML allows multimedia to be incorporated into the Internet by providing hyperlinking and the ability to use sound, animation, video, and graphics on a Web page.

The second event was the commercialization of the Internet. Companies saw an opportunity to not only communicate with new and potential customers,

Figure 11.2

A menu structure used to access information on the Internet

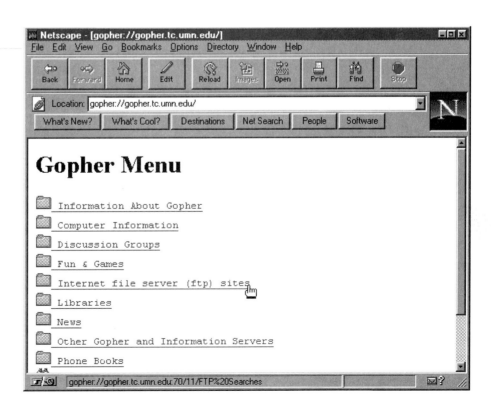

Figure 11.3
A Web page

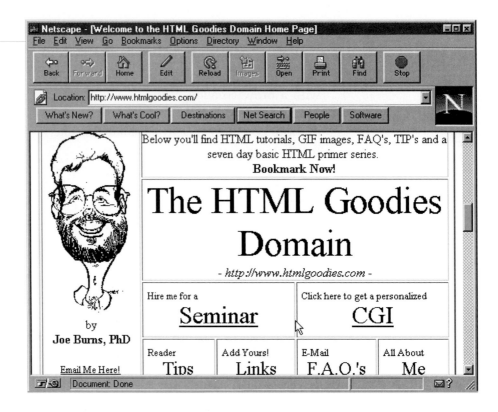

but actually sell their products through the Internet. When a new opportunity such as the WWW draws the attention of business, there is an enormous amount of money directed toward creating, in this case, an entire industry. New companies emerged and existing companies focused their efforts to compete as entirely new businesses and specialties opened up. The following are just some of the unique job descriptions and industries that arose as a result of the commercialization of the Internet:

- **Software developers** The development of software included creating browsers such as Netscape Navigator and Microsoft Internet Explorer, which are used to interpret and display HTML documents; creating development tools such as PageMill (and Page Site) and Front Page, which are used to create HTML documents and manage Web sites; and creating utilities to allow current applications to run over the Internet, for example, Shockwave, which allows Macromedia Director movies to be delivered via the Web.

- **Service providers** These companies provide individuals as well as businesses and organizations access to the Internet and its World Wide Web component.

- **Communications companies** This industry focuses on establishing the infrastructure necessary to provide Internet access throughout the world and to increase the speed of the network.

- **Web specialists** These people specialize in the development and support of Web sites. They include webmasters, network technicians, and multimedia developers focusing on creating material for the Web.

These two events provided the necessary structure for the development of the Web. But it was the ease at which a company, college, agency, or individual could become a global publisher that propelled the WWW's phenomenal growth. With very little training and only a small monthly fee, an individual could create, place, and maintain a home page—a Web page that serves as your main menu or home base—on the Web.

The shear number of Web sites poses a problem for companies, however. How can they attract individuals to their site, compel them to explore it, and have them return? One way is to provide the appropriate content—content that satisfies a need, answers a question, entertains, or provides something of value. Another way to draw visitors is to make the site compelling through the use of multimedia.

Multimedia on the World Wide Web

The World Wide Web part of the Internet has multimedia capabilities. Browsers such as Netscape Navigator and Microsoft Internet Explorer that interpret HTML documents allow graphics, sound, movies, and animation to be delivered to the user. Hypertext Markup Language allows developers to include hyperlinking in their Web documents, giving the user the ability to "navigate" from place to place in a document. An important advantage to developing multimedia applications for the WWW is that Web documents are created using HTML standards. Multimedia developers know that so long as their program complies with these standards, the user should be able to run the application. This is the case whether the user's computer runs on a Windows, Macintosh, or UNIX platform. Unfortunately, unlike the CD-ROM multimedia playback systems (such as MPC 3), there is no standard hardware configuration for computers linked to the Internet. For example, there may or may not be a sound card on a user's computer. This means that a multimedia title developed for delivery on a CD may not be appropriate for delivery on the Web.

LIMITATIONS OF THE INTERNET

There are significant considerations when developing applications for the Web, including file sizes and the playback system configuration (hardware and software). Most home computers are connected to the Internet through a modem and phone lines. A standard modem transfers data at only 28.8 kilobytes per second (28.8 Kbps) compared with a CD, which has transfer rates as high as 1,800 Kbps. Thus large files, especially video clips, sounds, and animations, can take an inordinately long time to move from a server to a home or office computer. There are techniques that can be used to speed up the transfer time and/or give the developer some control over the process:

- First, the transfer process can be completed more quickly by using *compression:* Files are compressed before they are sent to the user's computer and then decompressed as they are needed. In fact, the standard graphics file formats for the Web are GIF and JPEG, which are automatically compressed when they are created. **GIF** stands for Graphics Interchange Format and is the prevalent graphics format for images on the Web. **JPEG**, short for Joint Photographic Experts Group, the organization that created the standard, is used to digitize still photographic images.

- Second, a combination of CD and Internet delivery would allow the user to access large files from the CD and other information from the Web. An example would be a college class in which a CD is distributed to each student. The CD could contain video clips, 3-D animations, and other large files that are displayed more quickly from a CD than if they were delivered over the Web. As the student uses the CD, it automatically accesses up-to-date information from a Web site.

- Third, developers should consider using multimedia elements and development techniques that minimize file sizes. Such shortcuts include using animations instead of video clips, 2-D rather than 3-D animation, and 8-bit rather than 16-bit color and sound.

There are alternatives to using a modem with standard phone lines that are in use or under development. These include the following:

- **Cable modems** These utilize the coaxial television cable and provide speeds up to 100 times faster than modems using telephone wires.

- **ISDN** Short for Integrated Services Digital Network, **ISDN** utilizes dedicated switching equipment and existing telephone cables to provide speeds of up to 128 Kbps.

- **T1 lines** These are leased telephone lines that provide speeds of 1.5 Mbps.

- **T3 lines** These leased telephone lines provide speeds of 44.7 Mbps.

All of these techniques can provide faster transfer of data (at a cost), and companies and organizations use them to connect their intranets as well as to connect externally to the Internet. ***Intranets*** are internal organization networks that look and function like the Internet. A primary difference between the Internet and an intranet is that an intranet has restricted access—it may be limited to employee use only, for example. Intranets can allow access to the Internet, and persons outside a company such as customers and suppliers might be given access to a corporate intranet. Companies set up an intranet for a number of reasons:

- To provide a way for employees to collaborate on projects
- To provide access to company information such as sales or new product data
- To announce new policies or company events
- To deliver multimedia training materials

Because company networks are usually faster than the Internet in general, they can support multimedia-based training materials that are feasible for the intranets even though they would not be practical for the Internet. Until the bandwidth is increased to allow larger files to be sent over the general Internet more quickly, delivering multimedia titles via the Internet will remain a challenge.

DEVELOPING MULTIMEDIA FOR THE WORLD WIDE WEB

HTML is used to format the appearance of a document and to create links that allow the user to navigate throughout the document. Formatting, such as centering a heading or underlining a word, is done through the use of *tags*. Figure 11.4 shows two screens. The screen on the left is a Web page as it appears to a user. The screen on the right shows the tags used to format the text. Most tags are used in pairs. For example, the <Center> and </Center> pair causes whatever falls between them to be centered. The <B> and </B> tags cause whatever falls between them to be displayed in bold. The example in figure 11.4 has the following parts:

Figure 11.4

Two Web page screens—the user view and the tags

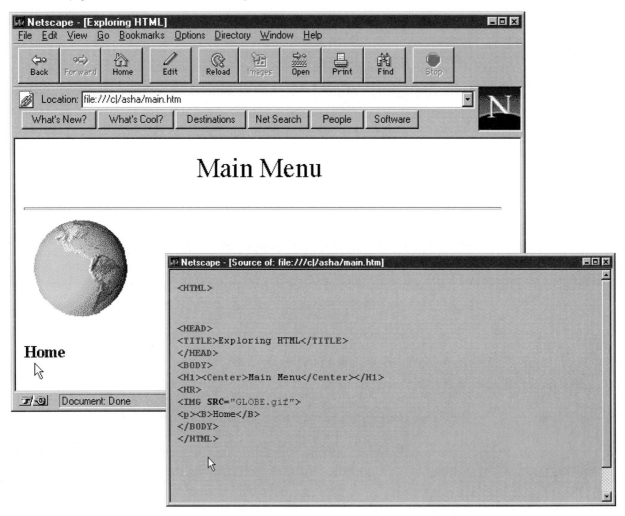

- A heading *Main Menu* that is centered by using the <Center> </Center> tags

- The word *Home* displayed in bold using the <B> </B> tags

Creating hyperlinks using HTML requires identifying the element (such as a word or graphic) that is to be linked and the destination for the link. Figure 11.5 shows a document with the word *Macromedia* identified as a hypertext word and the code that creates the link to another document. When the user clicks on the word, the linked document is displayed. HTML also allows a developer to easily insert graphics into a document. Figure 11.6 shows a document with a graphic and the code that specifies what graphic will be displayed.

Figure 11.5

A hypertext word and the code used to create the hyperlink

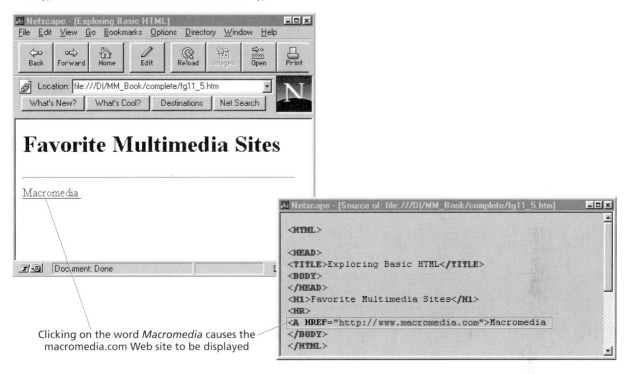

Clicking on the word *Macromedia* causes the macromedia.com Web site to be displayed

Figure 11.6

A graphic and the code used to specify which graphic to display

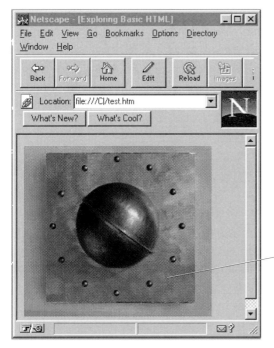
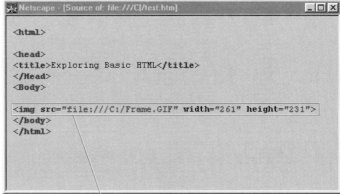

This code displays the graphic named *Frame.GIF*

Creating an HTML document is as easy as using a word processor to type the text with the formatting codes or using an HTML editor to insert the codes automatically. HTML has limited capabilities, however, and requires a programming language like Java to produce sophisticated features such as creating animations and searching a database. **Java**, developed by Sun Microsystems, has become popular for programming Web-related applications. It is a powerful language similar to C++ (widely used for developing various applications such as word processing programs and games) and is used to extend the functionality of HTML. For example, Java could be used to create an application that tracks stock prices and periodically displays them on a Web page. Such programs, called ***applets***, are developed for one specific purpose and like a template can be used in more than one application. Java applets are run within the browser on the user's computer and can be downloaded from a server. Applets are available from Sun Microsystems as well as third-party developers. Some are available free of charge and can be downloaded from the Web. Figure 11.7 shows part of a Java applet that causes an animation to play. Creating your own applets requires a high level of programming expertise, so you may want to determine if an applet is available that meets your needs before creating your own.

Another consideration when programming for the Web is the **Common Gateway Interface (CGI)**, a coding standard that allows programmers to write applications in various languages for the Web. CGI programs run on the server and provide a way for server-based programs to obtain information from a user and return information to the user. A typical application is one that allows the user to search a database. For example, a student may be studying the history of the Olympic Games and wants information on the gold medal winners of 1936. The student might use a CGI-based application that lets her search for the desired information. Figure 11.8 shows a Web page with space for the user to enter text for searching a database. The search application was created using CGI standards.

USING THE WEB AS A SOURCE OF MULTIMEDIA MATERIAL

The World Wide Web is an excellent source of clip art, sound and video clips, and photographs that can be used in a multimedia title. Virtually every company that provides these elements has a Web site and offers the ability to search its database for the desired material, pay for it, and often download it from the Web. Figure 11.9 shows the PhotoDisc Web site (http://www.photodisc.com). PhotoDisc has more than 50,000 digital images in categories such as People and Lifestyles; Science, Technology, and Medicine;

Figure 11.7

A Java applet that causes an animation to play

```java
//*********************************************************************
// ball.java
//*********************************************************************

import java.applet.*;
import java.awt.*;

//*********************************************************************
public class ball extends Applet implements Runnable
  {
    Graphics g;

    FontMetrics font_metrics;

    int string_width;
    int string_height;

    String message = "Hello World!";
    int point_size = 20;

    int message_length;
    char char_array[];

    int ball_size = 10;
    int delay_amount = 500;

    double decrement;

    int ball_x = 0;
    int ball_y = 0;

    //-----------------------------------------------------------------

    public void init()
      {
        g = getGraphics();

        String parameter;

        parameter = getParameter("MESSAGE");
        if (parameter != null)
          message = parameter;

        parameter = getParameter("POINT_SIZE");
        if (parameter != null)
          point_size = Integer.parseInt(parameter);

        parameter = getParameter("BALL_SIZE");
        if (parameter != null)
          ball_size = Integer.parseInt(parameter);

        parameter = getParameter("DELAY");
        if (parameter != null)
          delay_amount = Integer.parseInt(parameter);

        int divisions = 10;
        parameter = getParameter("DIVISIONS");
        if (parameter != null)
          divisions = Integer.parseInt(parameter);
```

Figure 11.8

A Web page with a search function

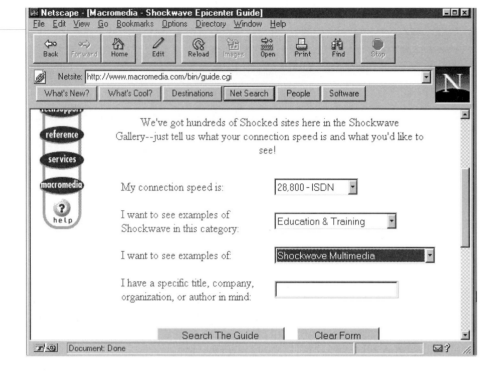

Figure 11.9

The PhotoDisc Web site

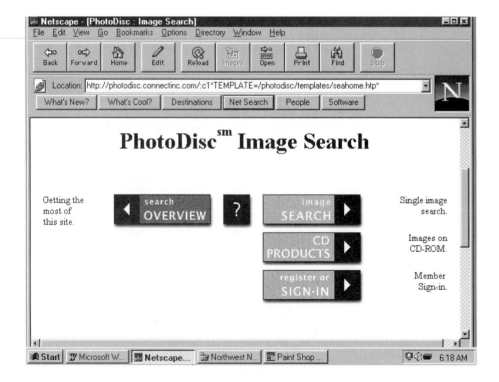

Nature, Wildlife, and the Environment; and Business and Occupations. The images, primarily photographs, are royalty free. You can search and download images as well as order from the CD-ROM collection and obtain a company catalog.

An example of a Web site for music and sound clips is the Multi-Media Music site (http://www.wavenet.com/~axgrindr/quimby.html) developed by Partners in Rhyme. The site provides high-quality audio for computer presentations and multimedia projects and includes sound utilities for the Mac and PC, sound effects, and (royalty-free) music loops. This site has links to other Web sites with sound clips that can be downloaded.

An example of a Web site with video clips is Four Palms (http://www.fourpalms.com). This site has royalty-free video clips in AVI, QuickTime, and MPEG formats.

Another way to locate multimedia materials is to search the Web using one or more of the search engines. Refining the search to obtain manageable results can be a challenge, however. For example, searching for "multimedia sounds" using the Excite search engine resulted in more than 40,000 hits.

Whenever the Web is used to obtain material such as sound clips or graphic images, care should be taken to ensure that copyright laws have not been violated. This can be done by dealing with well-known companies like PhotoDisc who control the copyright on the materials they distribute.

VIEWING MULTIMEDIA ON THE WEB

When developing multimedia for the Web, you need to understand how multimedia elements are viewed when delivered via the Internet. If you develop a multimedia title with sound, animation, and video, for example, and place the title on a CD, you could specify the configuration needed by the user (such as an MPC 3 system) and be fairly sure that the user could run the title. Web pages, however, are viewed with Web **browsers**, which are limited in their ability to automatically display certain files, including various image types as well as animations, video, and sound clips. For example, as mentioned in an earlier chapter, Macromedia Director is a popular program for creating animations. A Director "movie" can be saved in a format that allows it to be played directly from a CD. In order to play a Director movie from the Web, an additional program—Shockwave—needs to be installed on the user's computer.

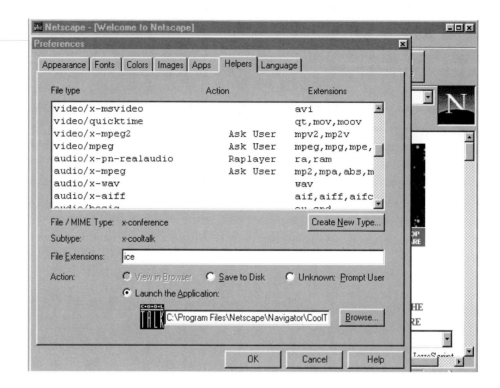

Figure 11.10

Netscape's Helpers Preferences dialog box

There are two types of programs that are used to view elements not viewable with a browser alone: helper applications and plug-ins.

Helper applications display an element (such as a video clip) in a separate window on the user's monitor. A ***plug-in*** (such as Shockwave) displays an element as though it were part of the browser. Because helper applications run independently, the user must download the application onto a hard drive and then configure the browser to use the application whenever a particular kind of file is encountered. Figure 11.10 shows Netscape's Helpers Preferences dialog box that lists the helper applications available to the browser.

ANIMATION ON THE WEB

Incorporating animation is an excellent way to increase the appeal of a Web site and help ensure return visits. Animations can be as simple as blinking text, marqueelike scrolling headlines, rotating logos, and 2-D action figures or as complex as 3-D virtual reality environments with user control. The following section introduces various types of animations and how they are created and displayed on the Web. As you read through this section, you can view a demonstration of these animations either from the CD or from the Web site.

interactive exercise

On the *Multimedia in Action* CD are some examples of Web-based animations.

1. Start the CD.
2. Choose Animation from the contents screen and read the instructions.
3. Click on the questions button and review the questions for Web animations.
4. Click on the demos button and choose Web animations.
5. After viewing the demo, respond to the questions previously reviewed.

These animations can also be viewed on the Web:

1. Launch your Web browser.
2. Display the Multimedia in Action Web site.
3. Follow the instructions to view the animations.

Animated text   Using the HTML <blink> command, you can cause text to flash on and off. To have the words YEAR-END SALE blink in the Web page document, you would include the following HTML code: <blink>YEAR-END SALE</blink>. Another way to animate text is by using a scrolling or marquee-type action to scroll text across the screen.

Animated GIF   The GIF graphics file format is a standard for the Web. GIFs are still images that can be combined to create an animation. A program called GIF Builder allows you to create an animation by displaying a series of GIF files. GIF Builder includes features for adjusting the speed of the animation and how many times (including continuously) it is played. This is an extremely easy way to create simple animations. Figure 11.11 shows the GIF Builder being used to create an animation.

Director movie   As mentioned earlier, a Director animation can be played using the Shockwave plug-in. This is a way to create somewhat sophisticated animations and have them delivered via the Web (see figure 11.12).

3-D environments   The computer language used to create 3-D environments on the Web that allow the user to move through a space or explore an object is called ***Virtual Reality Modeling Language (VRML)***. VRML technology is especially useful in creating games and educational titles. You need a browser that supports VRML or a plug-in to display VRML applications.

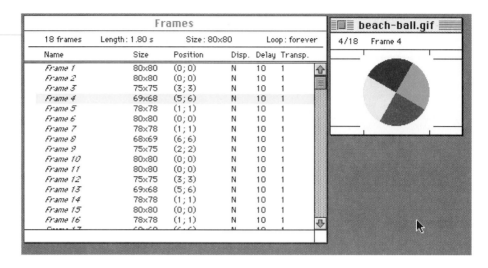

Figure 11.11
The GIF Builder being used to create an animation

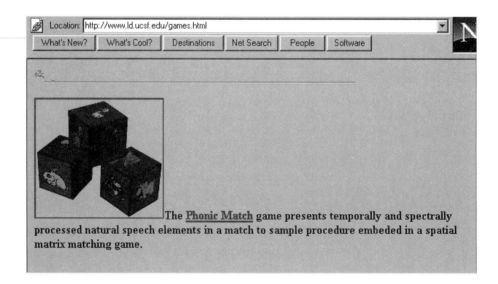

Figure 11.12
A Director movie being played using the Shockwave plug-in

Design Considerations for Multimedia on the Internet

LAYOUT AND FEATURES

The layout of a Web page, including the arrangement of the various elements and the features provided, is largely determined by the audience. Some users are interested primarily in getting to a site, quickly obtaining up-to-date information (such as stock quotes), and moving on. In this case, the design should be simple and straightforward. Other users are interested in the entertainment value of a Web site and may be willing to wait for large video or 3-D files to be displayed. For this reason, some companies provide Web sites that accommodate different audiences, thereby letting the user choose what best meets his or her needs.

As mentioned earlier in this chapter, a major consideration when developing multimedia that is delivered on the Web is the slow transfer speeds on the Internet. Transferring a 1 MB file across phone lines from a server to a user's computer can take more than five minutes. As CDs, disk drives, and computers in general have become faster in processing and/or transferring data, users have become less patient when having to wait for something to happen. Techniques such as keeping file sizes small and using file compression are useful. Other techniques include the following:

- Give the user control over whether or not to display or enlarge a graphic image. Figure 11.13 shows a Web page that has a small icon to represent a graphic. The icon loads much more quickly than the graphic itself. The user has the choice of clicking on the icon to display the graphic and waiting for it to display. A similar technique is to display a graphic in a thumbnail size and give the user the option of enlarging it (see figure 11.14).

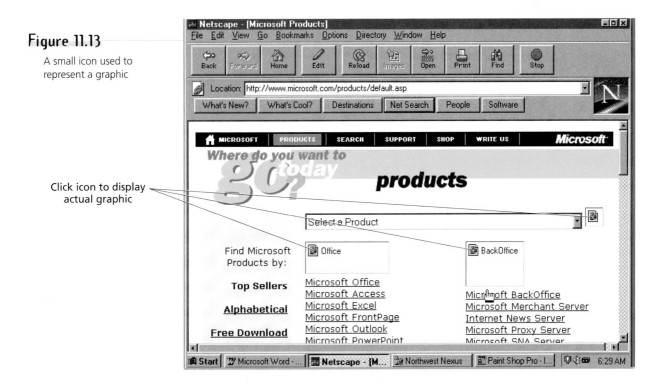

Figure 11.13

A small icon used to represent a graphic

Click icon to display actual graphic

Figure 11.14

Displaying a graphic in a thumbnail size

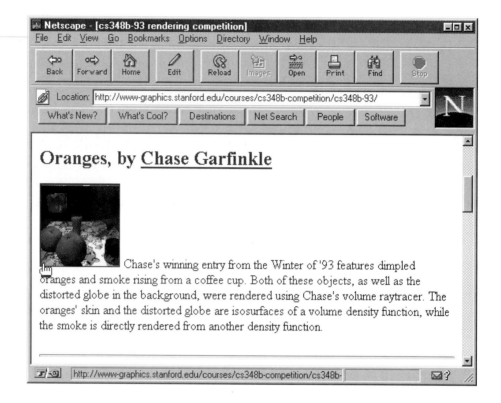

- Allow the user to be active while graphic images are being displayed. A graphic can be displayed in stages while the user is reading text, scrolling a page, or selecting a button hyperlinked to another Web page.

- Provide feedback to the user. Displaying the amount of time a file will take to download and showing a timer indicating the progress of the download helps the user decide whether or not to continue with the download and/or complete another task while waiting for the download to finish.

Other design considerations are the lack of standards for playback systems and not knowing what helper files and plug-ins the user has installed and what browser and version is being used. For example, the user may not have a plug-in or helper application that will play a certain video. Two common techniques are used to address these considerations. First, an alternative to the multimedia element could be provided. For example, an alternative to playing a video would be a series of still images and perhaps a text description of the action. Second, a button could be displayed that gives the user the option of downloading the necessary plug-in as shown in figure 11.15. This particular consideration will become less important as new versions of browsers come with plug-ins and as standards emerge.

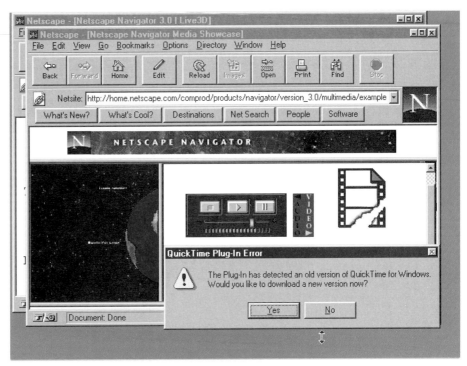

Figure 11.15

Displaying a button to allow a user to download a plug-in

HYPERLINKING

Some design considerations are related to the hyperlinking feature of a Web site. ***Web sites*** are made up of Web pages that are linked through some navigation scheme created by the site developer. A single Web site can contain an enormous amount of information, and the challenge to the designer is to provide a way for the user to find the necessary information and know where he or she is at any one time. In all sites there is a primary page often called the home page or welcome page. A ***home page*** is generally the anchor that is used as a reference point by the site developer: All other pages link to the home page, and usually there is a navigation button that takes the user to the home page from all of the other pages of the Web site (see figure 11.16). The navigation process may be complicated by the fact that the browser has a navigation scheme also. Figure 11.16 shows the Back and Forward buttons that are part of the browser window. Clicking on the Back button of the browser may not display the same Web page that clicking on the Back button in the application displays. This could be confusing to the user.

One of the values of the WWW is the number of Web sites and the vast amount of information available for free. It is a simple task to establish a link between one Web site and another. For example, an educational institution may find

Figure 11.16

A navigation button allowing the user to go to the home page

it beneficial to students to have a Web site that links to sites containing financial aid or employment information. Yet this is also one of the drawbacks of the WWW. It is relatively easy for an individual or organization to create a Web site, but there is no guarantee that the information is valid or current or that the site will be ongoing. Whenever a link to another site is provided, there is a risk that the user may start exploring the linked site and be unable or choose not to return to the original site.

One technique that helps a user always be able to quickly access your site is to have the user add the site to a list of favorites. Figure 11.17 shows a list (called Bookmarks) of favorite sites that can be accessed by simply clicking on the name.

To extend what you've learned, log on to the Internet at
http://www.thomson.com/wadsworth/shuman
You will find a wide variety of resources and activities related to this chapter.

Figure 11.17
A list of favorite sites

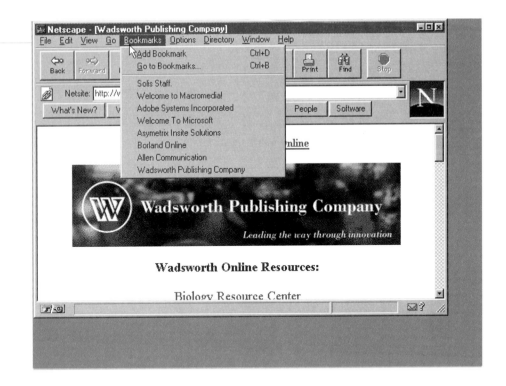

key terms

applet
browser
client
Common Gateway Interface (CGI)
compression
Graphics Interchange Format (GIF)
helper application
home page
Hypertext Markup Language (HTML)
Integrated Services Digital Network (ISDN)
Internet

intranet
Java
Joint Photographic Experts Group (JPEG)
network
plug-in
server
tags
Virtual Reality Modeling Language (VRML)
Web page
Web site
World Wide Web (WWW)

review questions

1. **T F** The Internet is a vast communications system linking computers around the world.

2. **T F** The World Wide Web was started by Macromedia.

3. The formatting standard used to create documents that can be viewed on the World Wide Web is called _____.

4. **T F** The commercialization of the Internet has been, in part, responsible for its significant growth.

5. Two standard graphics formats used on the Web are _____ and _____.

6. In HTML coding, the tags _____ and _____ would be used to turn on and off the bold feature.

7. _____ is a programming language specifically developed for the World Wide Web.

8. **T F** At this time it is not possible to view movie files on the Web.

9. Internal organization networks that look and function like the Internet are called _____.

10. **T F** Helper applications display an element, such as a video clip, as though it were part of the browser.

projects

1. Use a search engine to visit the following Web sites:

Apple	Microsoft
Asymetrix	MTV
Dorling Kindersley (DK)	Netscape
Edmark	PhotoDisc
Macromedia	Sun Microsystems

   Develop a report that includes the following:

   - How the company is involved in multimedia (products, services, education)
   - Who would be most likely to visit the site

- What value the site can offer the user
- What external links are provided and why
- What techniques are used to address the slow transfer speeds of the Internet
- Which sites are most effective from a visual design standpoint and why
- Which sites are most effective from a navigational design standpoint and why
- Which sites you feel are the most useful and why
- Which sites you feel are the least useful and why

Prepare a brief oral presentation of your report and be ready to present it to your class.

2. Search for Web sites that provide information about companies, products, or organizations related to multimedia and the Internet. For sites other than those in project 1 above, develop a report that includes the following:

- The site name and URL
- How the site relates to multimedia (products, services, education)
- Who would be most likely to visit the site
- Why you feel the site is useful
- What external links are provided and why
- What techniques are used to address the slow transfer speeds of the Internet

Prepare a brief oral presentation of your report and be ready to present it to your class.

Issues and Trends in Multimedia

AFTER COMPLETING THIS CHAPTER YOU WILL BE ABLE TO:

- Discuss the copyright issues affecting multimedia
- Identify the privacy issues in multimedia
- Define the censorship issues concerning multimedia
- Describe trends in the multimedia industry

Copyright Issues

The laws of *copyright* are designed to protect intellectual property rights and provide potential monetary rewards for inventiveness and hard work. In this way they foster creativity. However, the ease with which material can be copied, digitized, manipulated, incorporated into a title, and delivered to a mass market has prompted concern about the adequacy of existing copyright laws as they apply to the multimedia industry. Consider the following:

- Videotapes and videodiscs can be rented and played on a VCR or videodisc player linked to a computer, allowing the images to be captured and digitized

- TV programs can be captured from a television connected to a computer

- Scanners are used to digitize printed material, photographs, and slides

- Computers equipped with audio cards can capture and digitize music and other sounds from CDs and audiocassettes

Once in digital form, any number of changes can be made to images and sounds. The *Mona Lisa* can be given blue eyes, a Kenny G sound clip could be synchronized with Mickey Mouse playing the saxophone, and the original *King Kong* movie could be changed to show him climbing the Eiffel Tower instead of the Empire State Building. All this can be done from the desktop.

Business models providing licensing agreements and royalty fees have been in place for many years to provide protection and payment to those in the music, movie, and publishing industries. Providing the worldwide digital and transmission rights to a song, a work of art, or a story, however, has been approached cautiously. Copyright holders are leery of how multimedia developers can so easily manipulate digitized material and how quickly and widely it can be distributed, especially over the Internet. One example involved Microsoft founder Bill Gates who, in the late 1980s, formed Interactive Home Systems (IHS). The company initially tried to acquire the exclusive digital rights to certain artwork so that it could be used in various multimedia titles. Selling exclusive rights was contrary to how museums traditionally permitted use of their collections. IHS eventually settled for nonexclusive rights to more than 25,000 pieces of fine art from several collections. The artwork has been used as content in museum kiosks and multimedia CD-ROM titles such as Art Gallery, which has sold more than 100,000 copies.

There are several options available to a multimedia developer for obtaining content, including acquiring rights to copyrighted material, utilizing noncopyrighted or public-domain material, creating the material in-house, or contracting for original material.

ACQUIRING RIGHTS TO COPYRIGHTED MATERIAL

Simply purchasing a videotape or music CD does not authorize the buyer to copy a video or sound clip. No rights beyond personal use are provided. In order to use copyrighted material, a developer must determine which rights are needed. This can be complicated. For example, obtaining the right to use a video clip does not necessarily mean you have the right to use the music or a particular character in the video, nor does it necessarily mean that you can use the video clip in both a CD-ROM and on the Internet. These may require separate licensing agreements. In addition, there may be restrictions on granting permission to use copyrighted material, such as for nonprofit applications only.

After determining which rights are needed, the developer must identify who has the ability to convey them. In the music industry, a songwriter may hold the copyright, but it is usually administered by a music publisher, who in turn may employ an agency to negotiate and collect license fees which can range from $0.12 to $0.30 per title. Figure 12.1 lists organizations and agencies involved in acquiring rights to use copyrighted material in the music industry. Despite these organizations it is sometimes necessary to locate an individual who holds the copyright (author, artist, singer, or photographer) and negotiate the desired rights. This can be a time-consuming and expensive process.

Dozens of companies and various government agencies are involved in providing stock footage from film and video archives (see figure 12.2). Some companies specialize in particular content, such as Warren Miller Films (sports), or particular locations, such as Idea Factory (Las Vegas). Stock footage companies will work with a developer to search for the appropriate clips, clear the copyrights, and transfer the footage to the desired medium. Fees are usually based on a per-second charge ($5 to $75) with a minimum of 10 seconds. Additional fees may be charged for research, licensing, and duplication. Typically, rights are granted for use of the material for a particular project (title) and length of time (four to ten years). Although the multimedia industry uses relatively small amounts of film and video, many stock footage companies realize that multimedia developers may eventually comprise the largest consumers of their services. They are therefore willing to negotiate with developers on different ways to provide payments for their services, including receiving royalties on sales of each CD-ROM title.

USING MATERIAL IN THE PUBLIC DOMAIN

Materials that have no copyright are said to be in the **public domain** and can be used without permission. Either no copyright was issued (such as with certain government-generated materials), the copyright has expired (for

Figure 12.1

Organizations involved in granting rights to copyrighted music

ASCAP
7920 Sunset Blvd., Suite 300, Los Angeles, CA 90046
(213) 883-1000

BMI
320 W. 57th St., New York, NY 10019
(212) 586-2000

BZ/Rights and Permissions Inc.
125 W. 72d St., New York, NY 10023
(212) 580-0651

Copyright Music and Visuals
67 Portland St., Toronto, Ontario, Canada M5V 2M9
(416) 979-3333

The Harry Fox Agency Inc.
711 3d Ave., 8th Floor, New York, NY 10017
(212) 370-5330

SESAC
421 W. 54th St., New York, NY 10019
(212) 586-3450

The Winogradsky Co.
12408 Jolette Ave., Granada Hills, CA 91344
(818) 368-3538

2010 Media (Rights and Clearance Services)
http://www.2010media.com/produc/music/rights.htm
(818) 952-3651

some works this is 75 years after publication), or it was not renewed. There may be legal considerations when using public-domain materials, especially those related to derivative works, trademarks, and people.

Derivative works are those based on an original work, such as translations, abridgments, adaptations, or dramatizations. A ***trademark*** is a name, symbol, or other device identifying a product; it is officially registered with the U.S. government, and its use is legally restricted to its owner or manufacturer.

Figure 12.2

Organizations involved in providing stock footage from film and video archives

ABCNEWS VideoSource
125 West End Ave., New York, NY 10023
(800) 789-1250 or (212) 456-5421

Archive Films
530 W. 25th Street, New York, NY 10001
(800) 876-5115 or (212) 620-3955

CNN ImageSource
One CNN Center
North Tower, 4th Floor, Atlanta, GA 30303
(404) 827-1335

Creative Video
1465 Northride Dr., Ste. 110, Atlanta, GA 30318
(404) 355-5800

Fabulous Footage Inc.
19 Mercer St., Toronto, ON M5V 1H2, Canada
(800) 361-3456

2010 Media (Stock Film Footage)
http://www.2010media.com/produc/library/film.htm
(818) 952-3651

Trademark protection covers the title of a publishable work and, in the case of fiction, often the name of its characters. The rights of individuals include what's known as the **Right of Publicity**; this is a legal basis for requiring permission and/or payment for using a person's name, image, or persona. Consider the following examples.

The original story of Hansel and Gretel is more than 75 years old, in the public domain, and theoretically available for inclusion in a multimedia title; however, if the story has been changed or repurposed using another medium, such as a movie or play, a copyright might apply. Also, although a story such as *The Hounds of the Baskervilles* may be in the public domain, the character Sherlock Holmes may have been trademarked and, if so, could not be used without permission. Furthermore, the Right of Publicity precludes you from using the likeness of an individual—alive or dead. This would include a photo, drawing, or video clip of Elvis Presley or Marilyn Monroe or even Bill Clinton.

Copyright and trademark issues are important to multimedia developers and often call for the professional services of a lawyer.

Privacy Issues

In 1994 a CD-ROM was distributed that contained more than 2 million Oregon driver's license records, including each person's vital statistics and Social Security number. The State of Oregon had made the data available to the public (as do many other states), and an individual used it to create the CD-ROM title. Marketing departments from any number of companies would be interested in obtaining information provided on a driver's license, such as name, address, birth date, and gender. Although this may not be a technical violation of privacy laws, it does point out how easily personal data can be obtained and distributed.

Laws dealing with **privacy** include two issues that are important to multimedia developers. First, revealing embarrassing facts about an individual that would be considered offensive to a reasonable person and where there is no sufficient cause for the disclosure (such as a legitimate news story) may violate privacy laws. For example, revealing that an individual had a substance abuse problem 20 years earlier may constitute an invasion of privacy. Second, placing a person in a false light which causes undue stress on the individual could also constitute a violation of privacy. For example, showing a video clip of a woman walking in front of an adults-only theater on a dark, deserted street may imply she is a prostitute, when in fact she may have merely lost her way in an unfamiliar city. A multimedia developer therefore needs to be careful when revealing facts about a person and in how an individual is portrayed.

Censorship Issues

Controversy over the content of CD-ROM titles is essentially no different than other media. Pornography, violence, and racism are as much a concern in multimedia as in television, movies, and music. This is especially true of titles, such as games, that may be directed toward children. Should **censorship**—the official and authoritative examination, and possible expurgation, of material for appropriateness of content—be applied to multimedia titles? If so, the question becomes: Who will control the content—the multimedia industry or the government? The movie and music industries have rating systems and labeling to provide the consumer with an indication of the appropriateness of the content for various audiences. Potential consumers of CD-ROM titles usually have little more than the company's promotional material on the package to inform them of the content and its appropriateness for a particular audience. In some cases, the distribution channel has determined the suitability of a title. One retail chain decided not to carry a popular multimedia title that it considered to be too violent.

Concern over content can even lead to disputes between partners in a publishing arrangement. In 1994 the Voyager Company (a multimedia developer) and Apple Computer entered into an agreement in which Apple included Voyager's award-winning title Who Built America in a bundled software package distributed to K–12 schools and libraries. In late 1994, after distributing thousands of copies of the CD-ROM, Apple started receiving complaints about the incorporation of such topics as abortion and homosexuality. Apple reportedly asked Voyager to remove the objectionable topics and Voyager refused. Apple saw the issue as responding to customer concerns, whereas Voyager saw it as censorship. After several months of discussions, Apple and Voyager agreed that the title would not be distributed to elementary schools, but would continue to be made available to high schools.

The issue of censorship in multimedia will intensify as titles become more prevalent on the Internet. The Internet provides the opportunity for developers to deliver their titles to users without the need of a publisher, distributor, retailer, or other intermediary. Being "socially responsible"—however it is defined—will become more and more the purview of the multimedia developer.

Trends in the Multimedia Industry

This section discusses the major trends in the multimedia industry, including the delivery of titles over the Internet, advances in hardware, the emphasis on quality and storytelling, and the focus on non-consumer titles.

THE INTERNET

The Internet is having a dramatic effect on the delivery of multimedia titles. The rush to the Internet has caused some people to predict the demise of the CD-ROM. The CD-ROM has a limited storage capacity and its content is unchangeable, whereas the Internet promises to provide virtually unrestricted, easily updatable information. Even though multimedia elements can be conveyed over the Internet, a major problem is the slow speed for delivering large files—especially video, sound, and animation. The connection from the server containing the title to the home or office computer does not provide enough bandwidth to accommodate the large files. **Bandwidth** is the capacity of a device to process or transmit information; the more information it can handle per second, the greater its bandwidth. The technology to increase the bandwidth, including cable modems and ISDN lines, is available; the problem is the costs involved in deploying this technology. Because of this, the CD-ROM will be a primary delivery medium for multimedia titles for several years to come.

HARDWARE

Multimedia processor A new generation of processor chips that include *Multimedia Extensions (MMX)* technology has been developed by Intel to increase the performance of computer video, audio, communications, and graphics. The technology allows the processor to work on different data elements at once, increasing the overall work the processor can do. This results in richer colors, more vivid sounds, and smoother animations and video. The chip will be able to handle such tasks as graphics acceleration, sound, and video decompression without add-on cards. The processor will take advantage of Microsoft's DirectX technologies, which include DirectSound, DirectDraw, DirectInput, Direct3D, DirectMIDI, DirectMPEG, and DirectPlay. This could improve the performance of multimedia operations by as much as 400 percent. Hardware and software manufacturers, as well as multimedia developers and users, will benefit from a single multimedia standard supported by Intel and Microsoft. Especially promising are applications that combine the advantages of high-performance desktop computer processing and storage with the benefits of an Internet link for accessing multimedia on the Web.

DVD The *digital video disc*, or *DVD*, represents a major advance in CD technology. The DVD dramatically increases the capacity of CDs from 650 MB (million bytes) to as much as 17 GB (billion bytes). This allows full-length movies with different audio tracks (to accommodate various languages), and even different versions of the same movie (PG, PG-13, R) to be available on one disc. The technology involves increasing the data density by reducing the size of the pits and lands and providing double-layered and double-sided discs. Initially, businesses and institutions that can afford the cost ($500 or more) of the DVD drives (see figure 12.3) and that have high-capacity needs, such as the need to archive data or provide storage for reference information, will benefit from this technology. As the cost of the drives decreases so that households can more easily afford them, the movie industry and the multimedia game developers who require large disc capacity (one multimedia adventure title originally shipped on six CD-ROMs) will benefit as well.

Figure 12.3
The digital video disc

CD-E Phillips Electronics has developed the **CD-E (compact disc–erasable)**—an erasable disc that allows a user to update information on the disc and free up disc space by deleting unneeded data. A CD-E drive will be able to read, write, and overwrite erasable discs. In addition, these drives will be capable of reading all existing CD formats, such as CD-ROMs and Photo CDs. Erasable CDs will be especially beneficial in multimedia development environments, as well as to those needing to exchange data, archive large amounts of data, and back up data stored on hard drives.

DEVELOPMENT

Target market Because of the intense competition in the consumer market, the emphasis for multimedia titles will focus more and more on the corporate market. Corporate training is a multibillion-dollar industry, and companies are realizing that multimedia training delivered in a lab setting, through a company network to the desktop, or to the factory floor using a kiosk can be more timely and cost-effective than classroom instruction. Corporate marketing, including CD-ROM or online catalogs, distribution of CD-ROM promotional titles, and multimedia-based presentations, is another area that will attract developers. Interactive promotions exceed $100 million per year and are expected to grow to $4 billion by the year 2000. Every major advertising agency and many specialized agencies have a multimedia department eager to tap into this growing market.

Content For those developers who choose to focus on the consumer market, the emphasis will need to be on quality and content. Consumers will demand the best-quality graphics, especially 3-D animation, sound, and video. This will force up product costs—already some of the most popular titles such as Sierra On-line's Phantasmagoria require several million dollars in development and marketing costs. Content will be critical, and there will be a shift away from merely repurposing content (such as putting a book on a CD-ROM). Creativity and storytelling, wherein the user can interact by taking the perspective of different characters and influence the story, will be more prevalent; and more titles will use live actors as the Hollywood storytellers become more influential in the multimedia industry.

Ultimately, the successful developers of non-consumer titles will be those who can help solve the on-demand training and marketing needs of corporations, and the successful consumer-title developers will be those who can create compelling experiences!

To extend what you've learned, log on to the Internet at

http://www.thomson.com/wadsworth/shuman

You will find a wide variety of resources and activities related to this chapter.

key terms

bandwidth
CD-E
censorship
copyright
derivative works
digital video disc (DVD)

Multimedia Extensions (MMX)
privacy
public domain
Right of Publicity
trademark

review questions

1. **T F** Privacy laws are designed to protect intellectual property rights and provide potential monetary reward for inventiveness and hard work.

2. **T F** Purchasing a movie on videotape invests the buyer with the right to use a part of the movie in a multimedia title so long as the video is acknowledged in the title's credit section.

3. **T F** If a story is more than 75 years old, it is considered in the public domain and any characters in the story can be used without permission of the trademark holders.

4. **T F** Placing a person in a false light that causes undue stress on the individual could violate privacy laws.

5. **T F** The issue of censorship in multimedia will intensify as titles become more prevalent on the Internet.

6. A problem with distributing multimedia over the Internet is that the communications from the server (computer) containing the title to the home computer does not provide enough _____.

7. The _____ may be able to increase the capacity of CDs from 650 MB to as much as 17 GB.

8. **T F** Because of the intense competition in the consumer market, the emphasis for multimedia titles will focus more and more on the corporate-training market.

9. **T F** Copyrighted works are those based on an original work, such as translations, abridgments, adaptations, or dramatizations.

10. **T F** Trademark protection covers the title of a publishable work and, in the case of fiction, often the name of its characters.

projects

1. Research one of the following topics. Using at least three sources (magazine articles, interviews, Web sites, and so forth), develop a report that includes the following:

    - Copyright issues and multimedia
    - Privacy issues and multimedia
    - Censorship issues and multimedia
    - Trends in multimedia related to software
    - Trends in multimedia related to hardware
    - Trends in multimedia related to the Internet
    - Trends in multimedia related to 3-D animation
    - Other trends in multimedia

    Prepare a brief oral presentation of your report and be ready to present it to your class.

2. Contact an organization involved in granting rights to copyrighted music or providing stock footage from film and video archives and develop a report that includes the following:

   - Name, address, and URL of the organization
   - Contact person, title, and function
   - Primary services provided
   - Prices for services
   - Sample contract
   - Examples (actual) of how their services were used by multimedia developers

   Prepare a brief oral presentation of your report and be ready to present it to your class.

Part VI

Hands-on

Tutorial

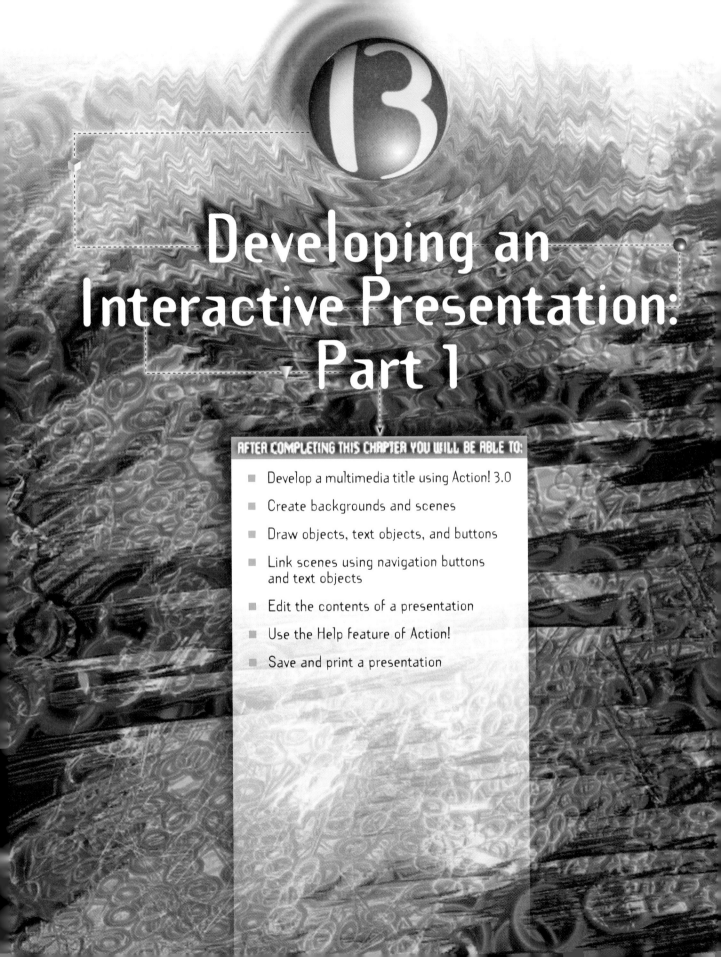

13

Developing an Interactive Presentation: Part 1

AFTER COMPLETING THIS CHAPTER YOU WILL BE ABLE TO:

- Develop a multimedia title using Action! 3.0
- Create backgrounds and scenes
- Draw objects, text objects, and buttons
- Link scenes using navigation buttons and text objects
- Edit the contents of a presentation
- Use the Help feature of Action!
- Save and print a presentation

What Is Action! 3.0?

Action! 3.0 is a program that allows you to develop multimedia presentations and interactive titles. Action! is made by Macromedia, the same company that makes Director, which is one of the most widely used multimedia authoring programs. Action! is easy to learn and straightforward to use when creating basic multimedia projects. It has characteristics of both a time-based program (like Director), which makes it easy to develop animations and apply special effects, and a card-based program (like HyperCard), which makes it easy to structure the information. Action! uses a series of scenes in which the various multimedia elements (text, sound, graphics) can enter and exit at any time. Figure 13.1 shows a thumbnail view of eight scenes that make up a multimedia title. Although these scenes are arranged in a linear fashion, they are hyperlinked, allowing the user to jump around randomly. Action! refers to the titles that are developed as *presentations*, because the program is useful for developing titles that can be used in a presentation setting. The presentations can also be user directed, however, and can be delivered on floppy disks and CDs, through kiosks, and even, with modification, over the Internet.

Figure 13.2 shows a scene with several multimedia elements and their properties.

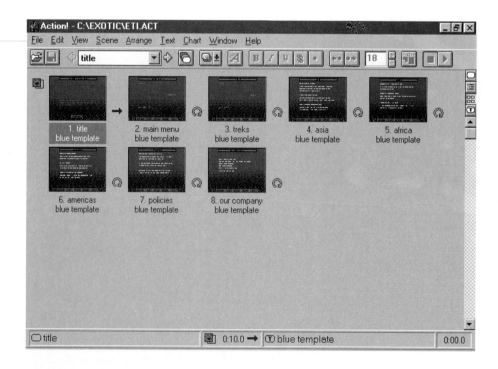

Figure 13.1

A thumbnail view of a presentation with eight scenes

Figure 13.2

An Action! scene with several multimedia elements

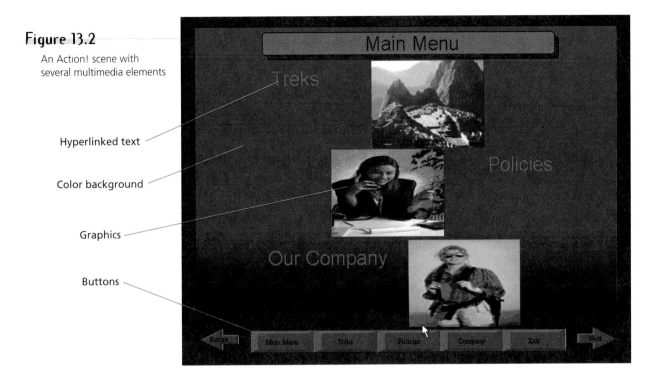

- Hyperlinked text
- Color background
- Graphics
- Buttons

In this chapter you learn how to use Action! 3.0 to develop a multimedia title. The chapter is written as a step-by-step tutorial that requires the Action! program to be installed on your computer. It is important that you follow each step carefully and understand each process before continuing to the next step.

In this chapter you are introduced to a fictitious company, Exotic Treks, which wants to develop a promotional multimedia title to give to potential customers that include travel agencies, clubs, and organizations. The title will have information about the company and the types of trips it organizes. It could be used as a presentation or as a stand-alone title distributed on CD or kiosk or delivered via the World Wide Web. Following is the planning that Exotic Treks has done for its title.

Project: Multimedia Informational Title

Objective: To provide our current and potential customers, and those influencing travel decisions, with information about who we are, what we do, and our corporate mission.

Audience: Those interested in unusual travel experiences (including individuals, clubs, and organizations) and those influencing travel decisions, such as travel industry personnel (travel agents, associations, outdoor equipment retailers).

Treatment: The title will have a simple, uncluttered, straightforward layout with an easy-to-understand navigation scheme. Elements will include text, animation, sound, video, and graphics, with an emphasis on pictures that represent our various trips.

Content Outline

Introduction: Purpose of the title

Examples of trips: Overview of selected treks

    Asia

    Africa

    Americas

Figure 13.3a

Storyboard for the Exotic Treks presentation

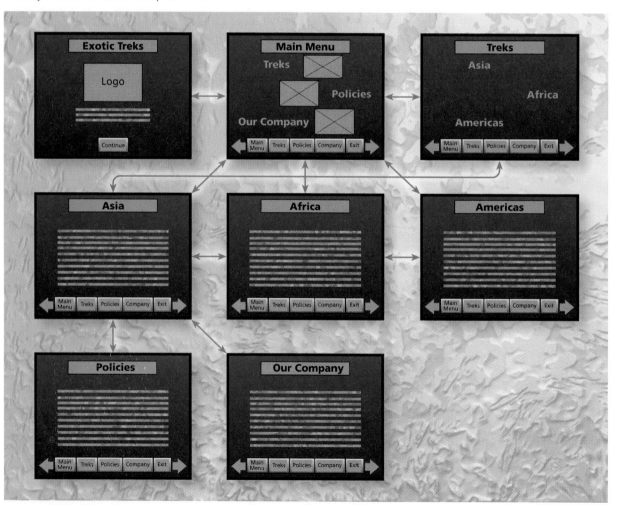

Policies
    Reservations
    Prices and dates
    Guarantee
Company information
    Mission
    Name
    Address
    Phone, fax, e-mail, Web site URL
Storyboard and flowchart

Figure 13.3 shows the storyboard and flowchart for the title.

Figure 13.3b
Flowchart for the Exotic Treks presentation

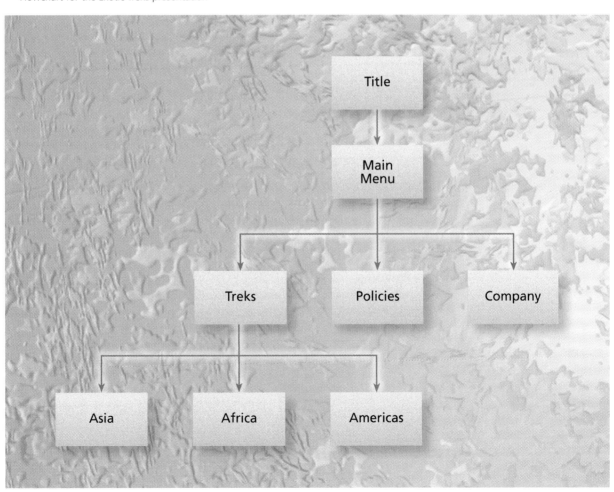

Developing the Exotic Treks Title

Working with the storyboard as a guide, you will use Action! 3.0 to create the Exotic Treks title. *Note:* This tutorial will work for the Windows 3.1 and Windows 95 environments. It is assumed that you have some basic experience with at least one of these systems.

1. Start Action! 3.0.

The Action! title screen appears with three options: Create a New Presentation, Open an Existing Presentation, or Exit Action!.

2. Click on Create a New Presentation.

A window appears, allowing you to choose a template for the new presentation. *Templates* are predesigned layouts that can save development time. They include a background color and/or graphic, and font sizes and types for headings and bulleted text. You will be developing your own template, so choose Blank.

3. Click on Blank.

4. Click on OK.

The Action! main screen appears, as shown in figure 13.4. Start by maximizing this window.

5. If necessary, enlarge the Action! window to its maximum size.

Refer to figure 13.4 as you read the following description of the parts of this screen.

Title bar The title bar indicates the filename of the presentation after it has been saved. Until it has been saved, the word *Untitled* appears.

Menu bar The menu bar lists the drop-down menus that contain the commands used to create a presentation. Take a moment to view each menu item. A brief description of the most commonly used commands is presented. The commands are explained more fully as they are used.

1. Click on File on the menu bar.

The File menu commands are similar to those in most applications. They allow you to open a new or old presentation and save, print, and play presentations.

2. Click on Edit on the menu bar.

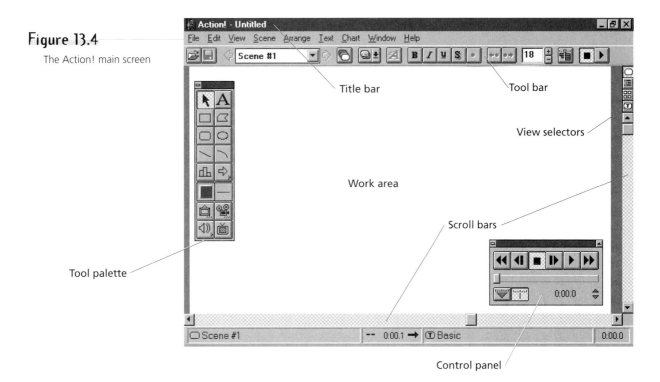

Figure 13.4
The Action! main screen

The Edit menu commands are also similar to other applications. You can copy, paste, and delete items. One of the most important commands is Undo. This command allows you to change your mind about an editing process or correct an unintentional operation. For example, you may have moved a picture on the screen and decide that the original position was better. You could use the Undo command to reverse your move. Not all operations can be undone. It is important to use the Undo command immediately after the operation you want undone.

3. Display the View menu.

The View menu allows you to specify what to view on the screen. For example, you can use it to view the template that has been selected for the presentation.

4. Display the Scene menu.

Action! uses scenes like a movie. You create scenes for different parts of a presentation. The Scene menu allows you to create new scenes and set scene options such as background colors and scene transitions.

5. Display the Arrange menu.

The Arrange commands can be used when working with objects such as pictures, text, and buttons. The commands let you group objects, and when they overlap, specify which ones will be in front.

6. Display the Text menu.

The Text menu is used to set the properties of text objects, such as the font, size, and alignment (left, center, right).

7. Display the Chart menu.

The Chart menu is used to create and edit charts for a presentation.

8. Display the Window menu.

The Window menu is used to change the working environment by allowing you to display or not display palettes, panels, and controls.

9. Display the Help menu.

The Help menu allows you to get help with the Action! program and is explained in depth later in this chapter.

Tool bar   The tool bar provides quick access to many of the commonly used features; its buttons correspond to various commands on the menu bar.

View selectors   These are used to change the view as you are developing a presentation. There are four views: scene, outline, scene sorter, and template.

Scroll bars   These are used to scroll the screen window up and down and left and right. In most cases, a scene will not be fully displayed as you are creating it. If you want to place an object such as a button near the bottom of the scene, for example, you would need to first scroll down.

Status bar   The status bar gives information about the current operation or the status of the program. If an object such as a picture is selected, the status bar displays information such as the picture's name and when it enters a scene. If no object is selected, the status bar provides scene information such as the name of the scene and its duration (length of time).

Work area   This part of the Action! screen is used to place and align the objects that make up various scenes and templates.

Working with Palettes and Panels

Action! provides several *palettes* and *panels* that offer additional tools and controls. These may or may not appear on your screen. If not, you can display them by selecting them from the Window menu. Figure 13.4 shows the tool palette and the control panel visible in the Action! window. The tool palette

Figure 13.5

The tool palette and control panel

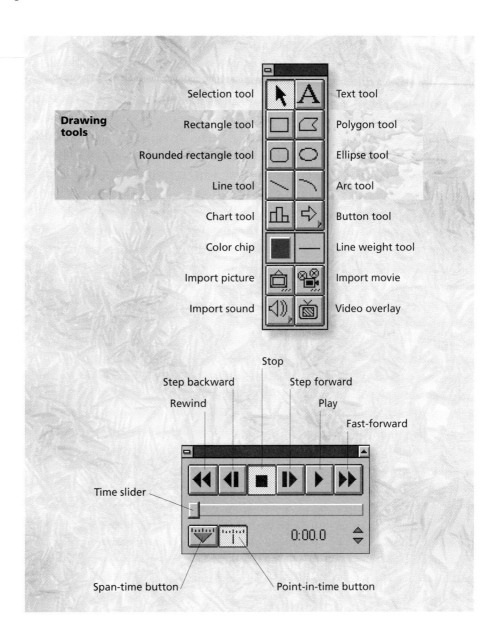

contains tools used to create objects within a scene. Figure 13.5 identifies the various tools. The tool palette can be used for the following tasks:

- Selecting objects for editing
- Drawing shapes such as lines, ovals, and rectangles
- Adding interactive buttons (for hyperlinking)
- Adding text, sounds, pictures, and movies
- Changing the color of objects and their line widths

Because there are so many features and tools available in Action!, you may need help in determining which tools do what. For many of the tools and buttons, you can display a description of the item on the status bar. The process is as follows:

- Point to the desired tool or button

- Hold down the mouse button (*Note:* Do not release the mouse button or the item will be activated)

- Read the description on the status bar

- Without releasing the mouse button, move the mouse pointer to a blank area and then release the mouse button

The process is shown in figure 13.6. Take a moment to try it.

1. Point to the text tool in the tool palette.

2. Hold down the mouse button.

3. Read the description on the status bar.

4. Point to a blank area of the screen.

Figure 13.6

Displaying a description of a tool

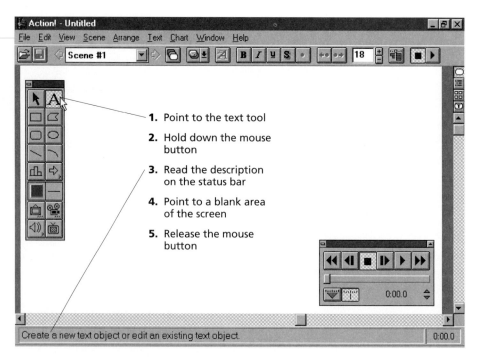

Chapter 13 *Developing an Interactive Multimedia Presentation: Part 1* **267**

5. Release the mouse button.

The control panel allows you to view a scene and move through it at a slow or fast speed. This is useful when testing a scene to determine how the various elements work together. For example, you can tell if a sound begins and ends at the correct time. Figure 13.5 labels the parts of the control panel. When developing a title, you may want as much of the work area as possible to be visible. You can turn on and off the display of the palettes and controls using the Window menu.

1. Click on the Window menu.

Notice that some of the items have a check mark next to them, indicating that they have been selected to view. Figure 13.7 shows the tool palette and the control panel having check marks. Clicking on a check mark will remove it from the menu and also remove the item from the work area. For now you will have only the tool palette displayed.

2. If necessary, click on Control Panel in the Window menu to remove the control panel from view.

3. If necessary, click on Tool Palette to display the tool palette in the work area.

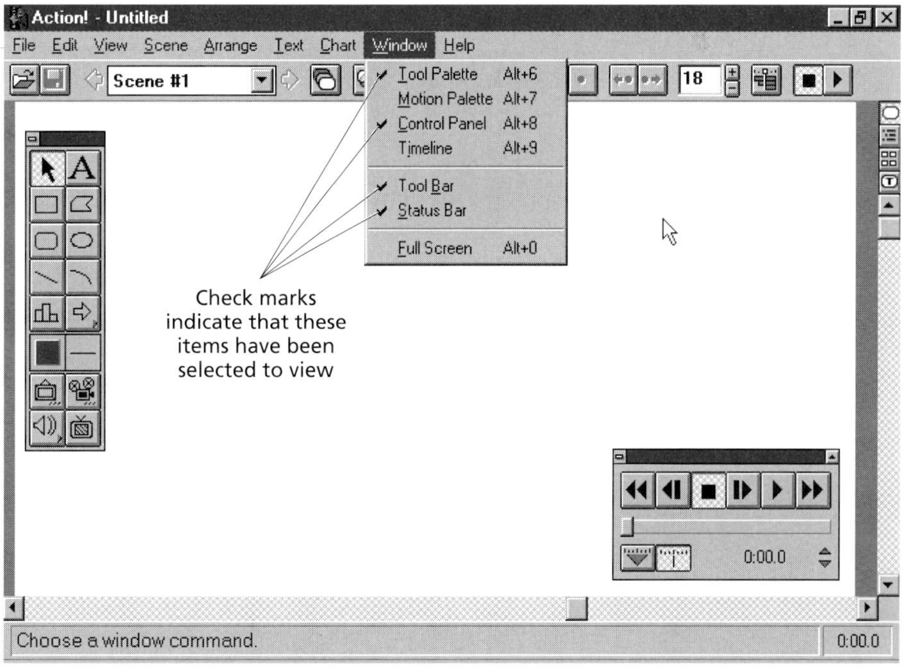

Figure 13.7

The Window menu indicating that the tool palette and control panel have been selected to view

Figure 13.8

Moving the tool palette

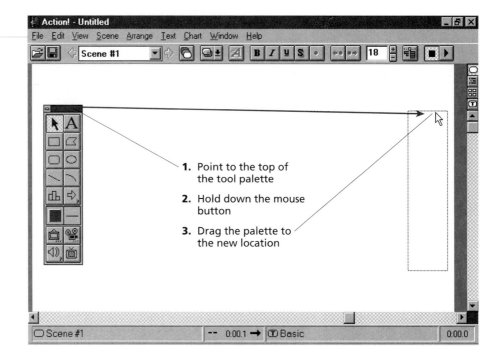

You can move the tool palette by pointing to the top bar, holding down the mouse button, dragging the tool palette to the desired location, and releasing the mouse button, as shown in figure 13.8.

4. Drag the tool palette to the location shown in figure 13.8.

As with many different programs, Action! often allows you to complete a task in more than one way. For example, there are four different ways to play a scene: using the Play command from the Scene menu; the play button on the tool bar; the control panel; and with the shortcut keys Ctrl-Shift-P. Which process you use is often a matter of personal preference. In this chapter alternative ways of completing a particular task are sometimes presented, but, if available, the menu process is usually shown.

Creating an Action! Scene

Before creating the first scene, you need to understand how to work with *layers*. Every scene can have three layers where *objects* (multimedia elements) are placed. There are three basic layers of objects in a scene: background, template, and scene.

Figure 13.9

The layers in a scene

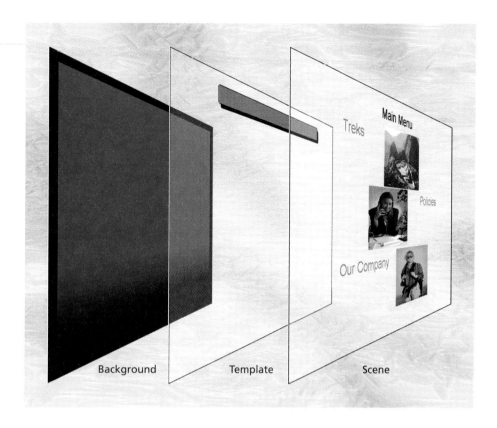

Background This contains only the background of the scene. You can specify color, patterns, and pictures to be placed on this layer.

Template Objects that are created in the template view occupy this layer. The template is used to save development time by placing here the elements that appear in more than one scene.

Scene Objects created in the scene view occupy this layer. Normally, this consists of elements that are unique to each scene.

Figure 13.9 show the relationships of these layers. For any particular scene, the background and template layers are optional. The storyboard can be used to suggest a template that might be applied to all or most of the scenes. In this case, there are two objects that are the same for each scene: the blue background and the heading box. Creating a template with these objects saves you the time of creating them for each scene and prevents the problem of inconsistency among scenes. For example, you would be assured that each scene had the same blue color and that the heading box was the same size and located in the same position in each scene.

Creating a Template

In this section you create and name the template for this presentation. Start by displaying the template view.

1. Choose Template from the View menu.

2. Choose New from the template options.

The template window appears. Notice that the words *Template #1* appear on the tool bar. Now specify a blue color for the background of the template.

3. Choose Background from the Template menu.

The Template Options dialog box appears with four tabs: Background, Time, Color, and Transition.

4. If necessary, click on the Background tab to display this option.

5. Click on Gradient.

Choosing Gradient will apply the selected color in a light-to-dark design. You can change the direction of the design using the buttons provided. The current design is from top to bottom. You can also change the color. Figure 13.10 shows the color palette that appears after you click on the color chip in the Colors box. Your color may already be set to blue.

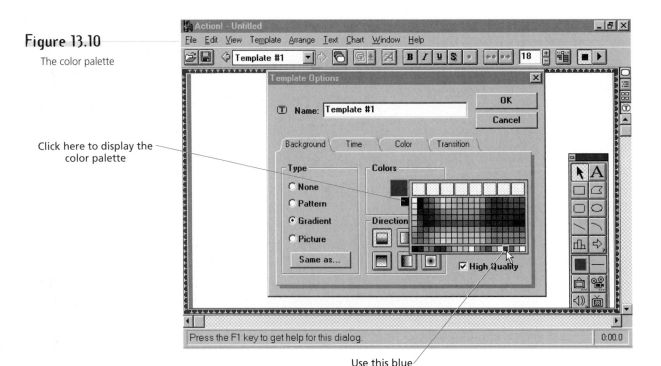

Figure 13.10
The color palette

Click here to display the color palette

Use this blue

6. If necessary, select the blue color from the color palette. (*Note:* Use the colors in the bottom row of the color palette.)

Before leaving this window, you will name the template *blue template*. Action! automatically provides generic names for objects, scenes, and templates, such as Template #1. It is important to label every object, scene, and template you create with a name that is relevant and meaningful. This makes it easier to work with these items as the size of the presentation grows.

7. Point to the *T* in the word *Template* in the Name field near the top of the Template Options dialog box.

8. Hold down the mouse button and drag the mouse to the right to select *Template #1*.

9. Type **blue template**

10. Click on OK to close the Template Options dialog box.

Next you will create the heading box that will contain the text heading for each scene. This is done with the rounded rectangle tool from the tool palette. Start by changing the color on the tool palette to pink.

11. Click on the color chip in the tool palette to display the color palette (see figure 13.11).

12. Click on the pink color in the bottom row of the color palette.

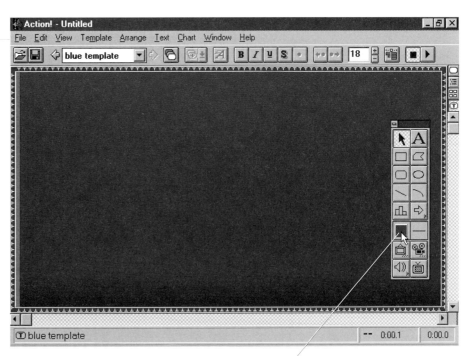

Figure 13.11

Click on the color chip to display the color palette

13. Click on the rounded rectangle tool in the tool palette. (see figure 13.12).

14. Draw the box shown in figure 13.12. Point to the upper-left corner of the box, hold down the mouse button, and drag the mouse pointer to the lower-right corner of the box. Then release the mouse button.

15. Click on the shadow button on the tool bar to give the heading box a shadow effect (see figure 13.12).

Now name the heading box object.

16. Point to the heading box and click the right mouse button to display a list of options.

17. Click on Motion.

The Object Options dialog box appears.

18. Change the name to **background heading box**

19. Click on OK.

This completes the template. Now return to the scene view.

20. Choose Scene from the View menu.

21. Choose Scene #1.

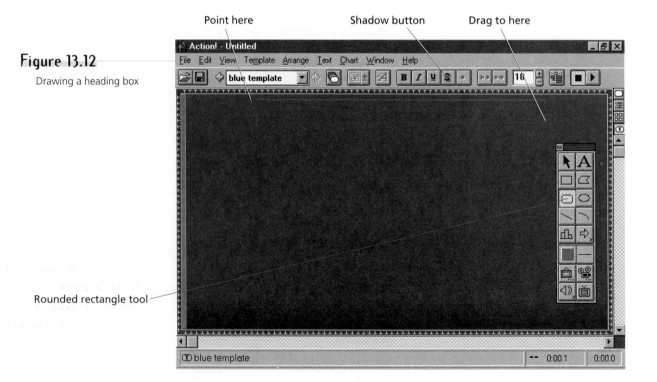

Figure 13.12
Drawing a heading box

Figure 13.13

The first scene

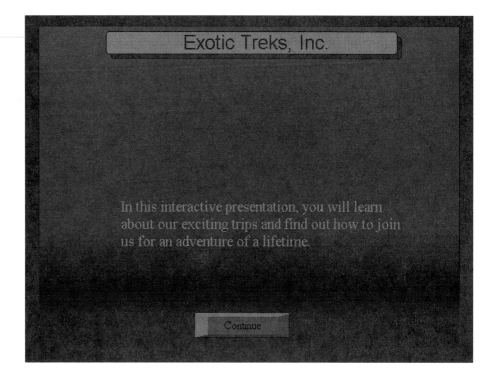

Figure 13.13 shows the first scene in the storyboard. The elements (objects) include the blue background, the heading box, text added to the heading box, body text, and a Continue button. The first step in creating a scene is to choose a template. In this case, you will specify the blue template.

1. Choose Use Template from the Scene menu.

2. Choose blue template.

The two objects, blue background and heading box, appear in the scene. It is important to understand that these appear in the scene but they are part of the template and can be changed only in the template view. Now you will use the Scene Options dialog box to specify a name for the scene and to specify how long the scene will run and what happens at the end of the scene.

3. Choose Options from the Scene menu.

4. Change the name Scene #1 to **title**

Now set the time for the scene. Time is displayed in the following format:

Figure 13.14

Setting the scene length to 10 seconds

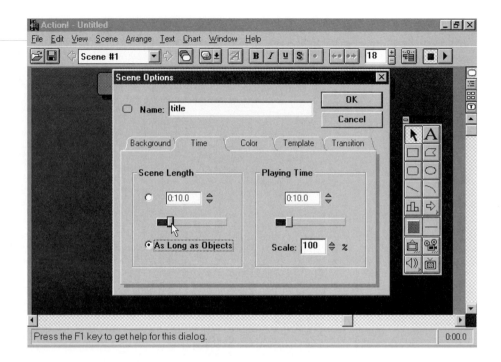

   5. Click on the Time tab.

Each scene has a specified length of time to be displayed. Objects within the scene can enter and exit at any time. This allows you to have various objects in motion or emphasized at any particular time. The Scene Length box is used to specify how long the scene will play. If you choose As Long as Objects, the scene will play until the last object leaves the scene. The Playing Time box allows you to speed up the playing of a scene without readjusting the time of each object. This first scene will have animated text, so you will select the As Long as Objects option. You can use the up and down arrows or the scroll bar to set times.

   6. If necessary, use the scroll bar or the arrows to set the Scene Length value to 10 seconds (see figure 13.14).

   7. If necessary, click on As Long as Objects to select it.

Your screen should resemble figure 13.14. Now specify what happens at the end of the scene.

   8. Click on the Transition tab.

There are two parts to this option: In Transition and At End of Scene. In Transition allows you to indicate the transition effect (such as Wipe Left) to

Chapter 13 Developing an Interactive Multimedia Presentation: Part 1 **275**

Figure 13.15

Setting the transition effect for the scene

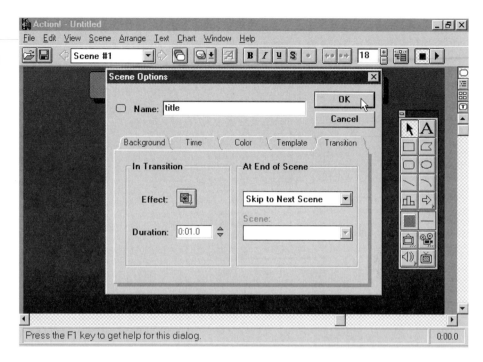

occur as the scene appears. At End of Scene allows you to indicate what happens when the scene ends. Start by setting the In Transition option.

9. Click on the button next to the word *Effect.*

10. Point to Wipe and click on Wipe Down.

Notice that the Effect button displays a graphic representing the selected effect.

11. Verify that the Skip to Next Scene option is displayed in the At End of Scene box.

Your screen should resemble figure 13.15. This completes the options for scene 1.

12. Click on OK.

Adding Text to a Scene

Next you will add the heading text object using the text tool and specify a motion for the text. Keep in mind as you work with a multimedia element, such as a text heading or a button, that it is considered an object. All objects have properties (such as their size, color, and the motion assigned to them). It is important that you understand how to create and edit an object and how to change its properties. Start by changing the color.

1. Change the color chip in the tool palette to black. (*Note:* Use the colors in the bottom row of the color palette.)

2. Click on the text tool.

3. Draw a text box just inside the heading box (see figure 13.16).

A window appears, allowing you to enter text and align it.

4. Type Exotic Treks, Inc.

5. Click on a blank area outside the text edit box.

The text appears on top of the heading box. You now have three objects in the scene: background, box, and text.

6. Click on the word *Treks* to select the text.

After selecting an object (as indicated by the handles surrounding it), you can copy, move, and delete it. You can also change its properties including its size and color. To move a selected object, you can point to it and, holding

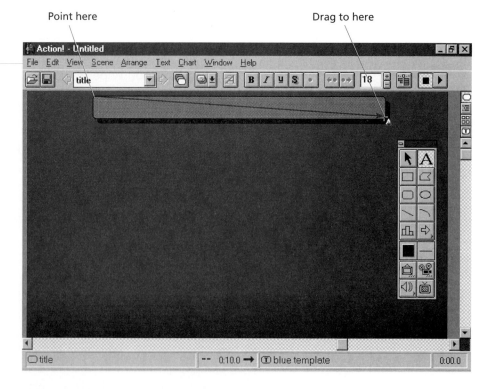

Figure 13.16

Drawing a text box just inside the heading box

Figure 13.17

The Font dialog box

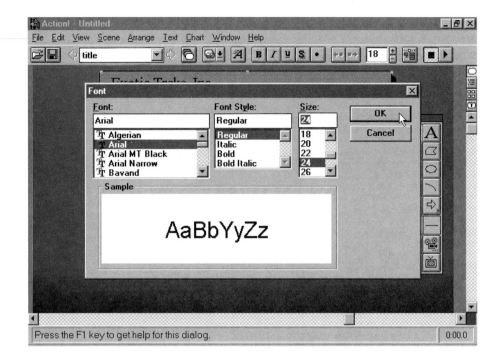

down the mouse button, drag it to a new location. To resize it you can point to a handle and, holding down the mouse button, drag the handle to a new location. In the case of a text object, you can change the font, type size, and style and align it within a text box. Complete the following steps to change the type to Arial size 24 and center-align the text.

7. With the text selected, choose Font from the Text menu.

The Font dialog box appears as shown in figure 13.17.

8. If necessary, choose the Arial font and size 24.

9. Click on OK to close the dialog box.

Now center-align the text within the text box handles using the Alignment option from the Text menu as shown in figure 13.18.

10. Choose Alignment from the Text menu.

11. Choose Center.

Now you will name the text object and cause it to enter the scene from the right side. This is done using the object motion options. To display these options, you can use the right mouse button.

Figure 13.18

The text alignment options

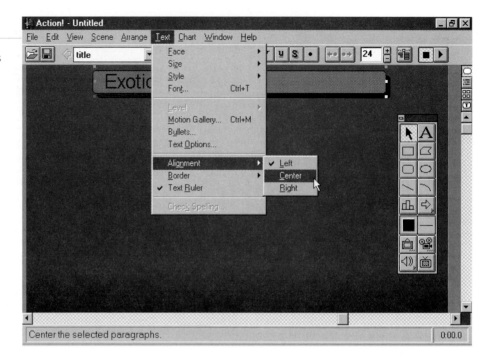

12. Point to the text and click the right mouse button.

A menu appears with several options.

13. Click on Motion (*not* Motion Gallery).

The Object Options dialog box appears. Start by renaming the object.

14. Change the object name from *Text #1* to **exotic treks**

Action! provides a number of options when applying motion to objects. Refer to figure 13.19 as you read the following descriptions of each motion option.

Start Time This is the time in the scene when the object appears. If it is set to 0, the object appears (or begins to appear if it is in motion) when the scene appears. If it is set to 0:02.0, the object appears 2 seconds after the scene appears.

Play Time This is how long the object stays visible in the scene. It is equal to the sum of the Enter, Hold, and Exit values.

To End of Scene Checking this option specifies that the object will remain visible until the end of the scene.

Figure 13.19

The object motion options

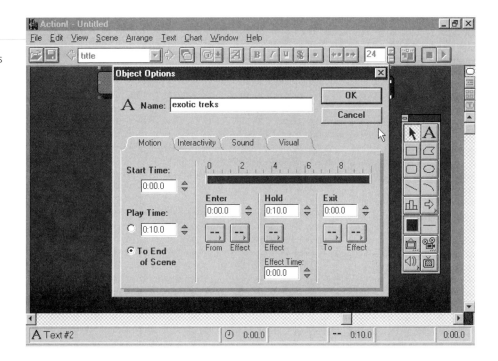

Every object has the following three time phases.

Enter This is used to specify how long an object takes to enter the scene and from which direction as well as what special effects are used. For example, you can have an object grow out from the center. If you specify 0:01.0 as the Enter value, the object will take 1 second to grow from the center and completely appear.

Hold This is used to specify how long an object remains motionless in a scene and what special effect is used. Special effects include sparkles and shimmers.

Exit This is used to specify how long an object takes to leave a scene and what effect is used as it does. For example, if the Exit value is set to 0:02.0 and the effect is set to Pull Right, the object will take 2 seconds to move right and leave the scene.

It is important to understand that the motion options allow you to specify which objects appear on the screen, how long they stay on the screen, and how they enter and exit a scene. This extremely powerful feature of Action! lets you create very compelling and interesting scenes.

Now use the motion options to specify how the text heading *Exotic Treks, Inc.* will enter the scene. In this case, you will have the object pulled in to the

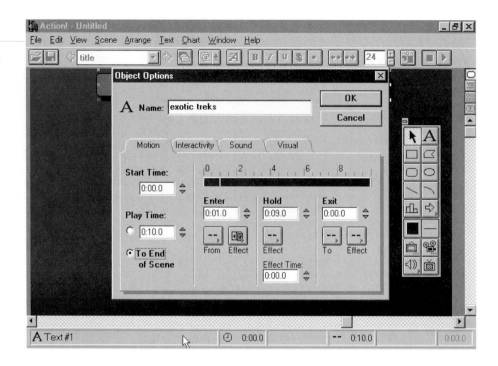

Figure 13.20
The completed dialog box

left and take 1 second for it to completely appear. The motion will begin when the scene appears, so you can leave the Start Time value at 0:00.0. The completed Object Options dialog box is shown in figure 13.20.

1. Make sure the Start Time value is set to 0:00.0.

2. Use the arrows to set the Enter value to 0:01.0.

3. Click on the Effect button for the Enter option.

Notice the different effects that can be used. Each of these has several variations. Take a moment to view them.

4. Point to Wipe and notice the variations.

5. Point to Pull and notice the variations.

6. Continue to view the variations for the other options. When done choose the Pull Left variation from the Pull option.

The Effect button now displays a graphic that represents the selected option (see figure 13.20).

7. Set the Play Time option to To End of Scene.

8. Click on OK to close the Object Options dialog box.

Now play the presentation to see how the text enters the scene.

Playing a Presentation or a Scene

Action allows you to play either the entire presentation or just one scene. Often when you are creating a presentation, you want to see how a single scene works. To view the presentation, you would choose Play Presentation from the File menu (or use the Ctrl-Enter keystroke combination). To play a single scene, you choose Play from the Scene menu (or press Ctrl-Shift-P). In either case, you can press Esc to stop the playback. Another useful way to play a scene is to use the control panel. The advantage of using the control panel is that it allows you to start, stop, pause, rewind, and step through the scene one-tenth of a second at a time (see figure 13.5). You will use the control panel to view the current scene as it would play in a presentation.

1. Choose Control Panel from the Window menu to display it.

2. Arrange the tool palette and the control panel so that the heading box is in view. (*Note:* You can move them by pointing to the top of each one, holding down the mouse button, and dragging to a new location.)

The control panel has two buttons that determine which objects are in view when the motion in a scene is stopped: span-time and point-in-time.

Span-time view This displays all objects in the scene in their hold positions. It is used to check the position of one object in relation to another, even when they do not appear in the scene at the same time.

Point-in-time view This displays the objects as they would appear at any given time. This is the way the user would see the objects.

3. Click on the point-in-time button.

4. Click on the rewind button.

5. Click on the play button.

The scene is played and the text is pulled to the left and completely appears after 1 second. Notice that the time slider ends at 10 seconds, the time set for the scene.

6. Click on the rewind button.

7. Remove the control panel from view.

Saving a Presentation

Before continuing you will save the presentation. It is recommended that you save early and often in the development of a presentation.

1. Choose Save As from the File menu.

The Save Presentation dialog box appears, allowing you to specify a filename and a location to save to. Currently, the filename is **.act*. The *act* is the filename extension for an Action! file. The location (directory or folder) depends on the setting for your computer. You will save the file using the name *eti.act*. Optionally, you can add your initials to the filename, such as *eti-js.act*. Also, you will specify the drive and folder of your choice.

2. Specify a filename for the presentation.

3. Specify a folder or directory for the file.

4. Click on OK.

Notice that the title bar now displays the name you specified when saving the presentation.

Adding a Scene

The next object that will be placed in the title scene is the Continue button that, when pressed, will allow the user to go to the next screen. Before adding a Continue button to the title scene, you will add a second scene to the presentation. Figure 13.21 shows scene 2, the Main Menu. This scene will use the same template as the title scene.

1. Choose New from the Scene menu.

The blue template is automatically applied to the new scene. Although there are several objects in this scene (the text heading, the three options, and the navigation buttons), for now you will insert only the text heading and set the scene options for this scene.

2. Verify that the color chip in the tool palette displays black.

3. Use the text tool in the tool palette to draw a text box and insert the words *Main Menu* as shown in Figure 13.21.

4. Change the font to Arial size 24 and center-align the words *Main Menu*. (*Note:* Use the same process as you did for centering the Exotic Trek text in the title scene.)

Figure 13.21

Scene 2, the main menu

5. Choose Options from the Scene menu.
6. Change the scene name to **main menu**
7. Click on the Time tab.
8. If necessary, set the scene length to 10 seconds and click on the As Long as Objects option.

Next you will set the transition for this scene. Because you want the user to control the presentation, you need a way to hold this scene. You can do this by replaying the scene. In essence, this causes the scene to start again after the scene ends in 10 seconds. If there is an animation or some effect, such as the heading text that is pulled into the scene, this is repeated. If there is no effect, the scene appears to remain static. The Replay Current Scene option can be used to cause the scene to play in a continuous loop.

9. Click on the Transition tab.
10. Click on the down arrow in the At End of Scene box.
11. Choose Replay Current Scene.
12. Click on OK.

Figure 13.22

Scrolling the window to display the bottom of the scene

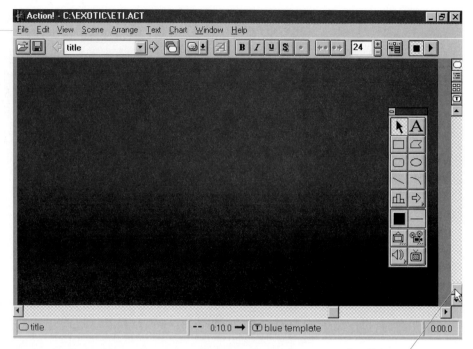

Scroll down

Now you are ready to link the first and second scenes. First, you will need to display scene 1. There are three easy ways to display a particular scene. You can drop down a list of the scenes and choose any scene to go to, you can use the arrows on the tool bar to go to an adjacent scene, or you can use the [Page Up] and [Page Down] keys to move to the previous or next scene.

13. Click on the words *main menu* on the tool bar to display the list of scenes and choose *title*.

14. Use the scroll bar at the right of the window to scroll down and display the bottom of the scene (see figure 13.22).

Creating Buttons and Linking Scenes

Action! provides an easy way for you to create buttons using the button tool in the tool palette. First, change the color.

1. Change the color chip in the tool palette to pink.

2. Click on the button tool in the tool palette.

A window appears with several button shapes.

3. Choose the button in the upper-left corner of the window (the square 3-D button).

Figure 13.23
Drawing the Continue button

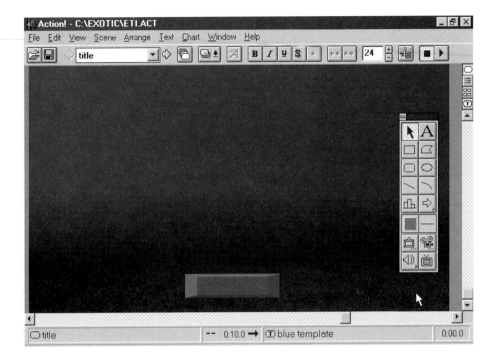

4. Draw the button as shown in figure 13.23.

Now you will set the options for the button, including the name, caption, and interactivity (that is, what happens when the user clicks on the button).

5. Point to the button and click the right mouse button.

6. Choose Button from the list.

The Object Options dialog box appears. You will use this dialog box to name the button *Continue* and to specify a caption. Notice the text field near the middle of the dialog box. Whatever you type in the text field will appear on the button. Start by changing the name of the button and then specifying a caption.

7. Change the name from Button #1 to **continue**

8. Type **Continue** in the Text field for the button caption.

9. Verify that the Position option is set to Center.

Now continue by specifying what happens when the user clicks on the button.

10. Click on the Interactivity tab.

The Interactivity panel appears, allowing you to specify what happens when the user clicks on the button. There are three parts to the panel.

Highlight Selecting this option causes the button to flash, giving the user an indication that the button has been clicked. This is important feedback because it prevents the user from continuing to click on the button, not realizing that the program has registered the action.

Action This is used to specify which scene or which part of the current scene the program will jump to.

Sound This option is used to specify whether a sound will be played when the button is clicked.

In this case, you want to display the main menu scene when the button is clicked.

11. Verify that the Highlight option is selected (it will appear checked).

12. Click on Action to select it (it will appear checked).

13. Click on the down arrow for the Action box and display the beginning of the list.

Take a moment to scroll through these options as you read a description of them:

Action	Description
Pause/Continue	Pauses the scene. Resume by pressing the spacebar, clicking the right mouse button, or clicking on the object.
Stop	Exits the presentation.
Go to (Next, Previous, First, Scene Start)	Jumps to the specified scene and has Action! remember the scene of origin. The Return option is used to return to the scene of origin.
Go to Link	Jumps to a specified scene. Not available in templates.
Return	Returns to the scene from which the most recent Go To was used.
Skip to (Next, Previous, First, Scene Start)	Jumps to the specified scene, but Action! does not remember the scene of origin.
Skip to Link	Jumps to a specified scene. Not available in templates.

Chapter 13 *Developing an Interactive Multimedia Presentation: Part 1* **287**

Because there will be a Return button available in most scenes, you will use the Go To options rather than the Skip To options. In this case, you will establish a link to the main menu scene.

14. Choose the Go to Link option.

15. Click on the down arrow in the Destination Scene box.

16. Choose main menu.

Your screen should resemble figure 13.24.

17. Click on OK to close the Object Options dialog box.

Study the Continue button. You may need to change the type size of the word *Continue* in order to have it fit within the button.

18. With the button selected, that is, the handles displayed, choose Size from the Text menu.

19. Choose 14.

Before testing the presentation, you need to specify a pause for the title scene. Currently, the title scene is set to play for 10 seconds, after which

Figure 13.24

The completed dialog box

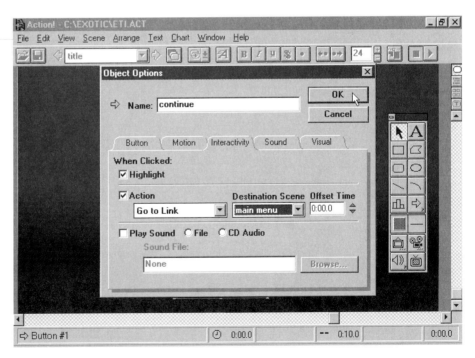

it will automatically go to the next scene. You could specify that the title scene continue to play in a loop, but this would mean that after 10 seconds it would start again. Thus, the text *(Exotic Treks, Inc.)* would be pulled in again. In some cases, this might be desirable; in this case, however, you want the title scene to appear, play, and then pause. You can cause a scene to pause by setting a time less than or equal to the scene length (10 seconds for the title scene).

20. Choose Add Pause from the Scene menu.

21. Set the Pause Time value to 9 seconds and click on OK.

Now test the interactivity by playing the presentation.

22. Choose Play Presentation from the File menu.

The title scene appears, the text is pulled to the left, and in 9 seconds the scene pauses. Notice that the pointer changes to . This indicates that the scene has paused.

23. Click on the Continue button.

The main menu scene appears. Now exit the presentation.

24. Press Esc to stop the presentation.

25. Save the presentation.

Copying Scenes

The next six scenes can be created quickly by copying the main menu scene and changing the headings. Copying scenes is done using the scene sorter view. You have already worked in the template view and the scene view. The scene sorter view shows thumbnails of the scenes as well as information such as their length of playing time, transition, and background.

1. Choose Scene Sorter from the View menu.

Figure 13.25 describes the parts of the scene sorter window. To copy a scene, you select the desired scene and choose Copy and then Paste from the Edit menu. You can verify that a scene is selected by the fact that its name and the template name are highlighted. An easier way to copy scenes is by using the new scene button on the tool bar. Clicking on this button will insert a new scene after the selected one. Only the scene template will be copied, however.

Chapter 13 Developing an Interactive Multimedia Presentation: Part 1   **289**

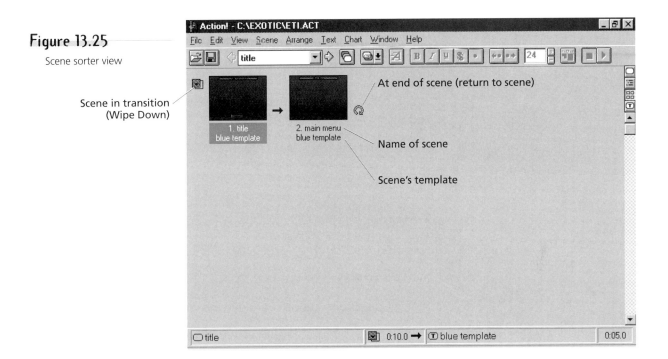

Figure 13.25
Scene sorter view

2. Verify that the main menu scene is selected and choose Copy from the Edit menu.

Before pasting the new scene, you must indicate where it will appear. If you paste it with a scene selected (in this case, main menu), it will be inserted *before* the selected scene. To have it appear *after* the main menu scene, you must first deselect the main menu scene.

3. Click on a blank area in the scene sorter window to deselect the main menu scene.

4. Verify that the main menu scene is deselected (the name should not be highlighted).

5. Choose Paste from the Edit menu.

A third scene, named Scene #1 because you have given names to the other two scenes, appears after the main menu scene. Notice that the blue template is used for the new scene.

6. Click on a blank area.

7. Choose Paste from the Edit menu.

8. Continue to choose Paste four more times to create a total of eight scenes. Be sure to click on a blank area after each paste operation.

Now edit each scene to add text for the heading. Start with Scene #1.

9. Choose Scene from the View menu and choose Scene #1 (the first copied scene).

10. Click twice on the words *Main Menu* to display the text edit box.

11. If necessary, drag the mouse pointer across the words *Main Menu* to select them.

12. Type **Treks** and click on a blank area to enter the text.

13. Choose Options from the Scene menu.

14. Change the name of the scene to **treks**

15. Verify that the Scene Length value in the Time panel is 10 seconds.

16. Verify that the At End of Scene option is set to Replay Current Scene in the Transition panel.

17. Click on OK to close the Scene Options dialog box.

18. Display Scene #2.

19. Continue to edit the remaining scenes to change the headings as follows:

Asia   Africa   Americas   Policies   Our Company

20. Change the name to match the heading

21. Verify that the Time and Transition options are set appropriately.

22. Save the presentation.

Creating the Navigation Bar

Now create the navigation bar for the main menu scene. This navigation bar will be duplicated for the remaining scenes. Start by creating the two arrow buttons. You will create the left-arrow button, copy it to ensure the same size, and then change it to a right-pointing arrow.

1. Display the main menu scene.

Figure 13.26

Drawing an arrow button

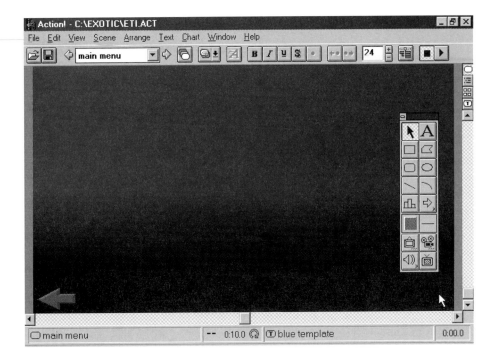

2. Use the scroll bar to display the bottom of the scene.

3. Verify that the color chip in the tool palette is pink.

4. Choose the button tool and select the 3-D left-arrow button.

5. Draw the button shown in figure 13.26. (*Note:* You can resize the button by dragging a handle.)

6. Point to the button, click the right mouse button, and choose Button.

7. Change the name to **return** and the caption to **Return**

8. Click on OK.

9. With the button selected, choose Size from the Text menu and change the size to 10.

Now create the other arrow button.

10. With the arrow button selected, choose Copy from the Edit menu.

11. Choose Paste from the Edit menu.

When you copy an object, the new object is pasted on top of the original one. To separate the two objects, drag the top one to another location.

Figure 13.27

Moving the copied arrow

12. Drag the copied button to the right side of the scene (see figure 13.27).
13. Click on the copied button using the right mouse button.
14. Choose Button.
15. Change the name to **next** and the caption to **Next**
16. Click on the 3-D right-pointing arrow.
17. Click on OK.

Now create the five rectangle buttons. To make sure that the buttons will fit across the screen, you will use the Visual panel to check the width of each button.

1. Choose the square 3-D button from the button tool palette.
2. Draw the button as shown in figure 13.28.
3. Right-click on the button and choose Visual.

The Visual panel allows you to, among other things, specify the size of an object.

Chapter 13 *Developing an Interactive Multimedia Presentation: Part 1*   **293**

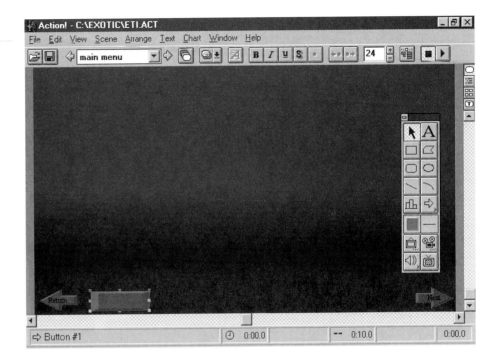

Figure 13.28
Drawing a rectangular button

4. Change the setting for the object to: Width 90 (pixels) and Height 30.

5. Click on OK.

6. With the button selected, choose Copy from the Edit menu.

7. Choose Paste from the Edit menu.

8. Move the newly copied button to the right as shown in figure 13.29.

9. With the copied button selected, use the arrow keys to align the copied button with the first button.

10. Continue to paste and reposition three more buttons.

Now align the buttons to resemble figure 13.30. This can best be done using the nudge buttons (the keyboard arrow keys).

11. Align the buttons to resemble figure 13.30.

Now name and specify a caption for each button.

12. Use the right mouse button to click on the leftmost rectangular button.

13. Select Button.

Figure 13.29

Moving the copied button

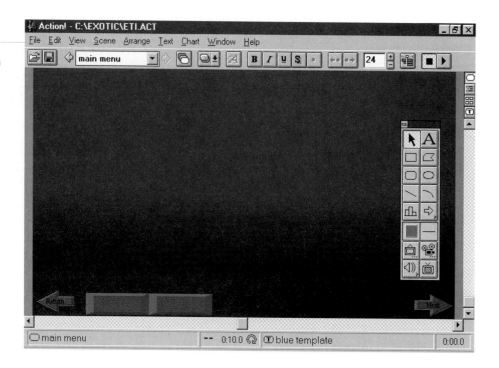

Figure 13.30

The buttons aligned

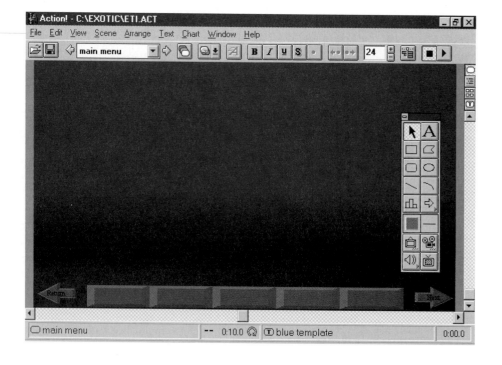

Chapter 13 *Developing an Interactive Multimedia Presentation: Part 1* **295**

Figure 13.31

The completed navigation buttons

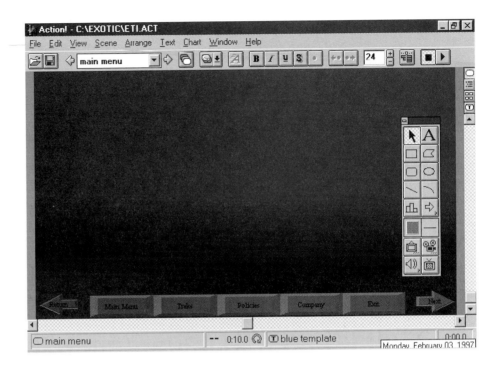

14. Change the name to **main menu** and the caption to **Main Menu**

15. Click on OK.

16. Change the text size to 10.

17. Continue to name and specify a caption for the other buttons as shown in figure 13.31.

Next, set the interactivity for each navigation button.

1. Right-click on the Return button.

2. Choose Interactivity.

3. Verify that the Highlight option is checked.

4. Turn on the Action option so that it appears checked.

5. Set the action to Return.

Now close the Object Options dialog box and continue with the other buttons.

6. Click on OK.

7. Right-click on the Main Menu 3-D rectangle button.

8. Display the Interactivity panel.

9. Activate the Highlight and Action options.

10. Select Go to Link in the Action box.

11. Select main menu in the Destination Scene box.

12. Click on OK.

13. Continue to specify the highlight, action, and destination for the remaining 3-D rectangle buttons. Set the action for the Exit button to Stop.

14. Set the action for the Next button to Go to Next.

Copying a Group of Objects

Currently, the navigation buttons are on only the main menu scene. You can copy them to the other scenes by grouping them and then using the Copy and Paste commands. To group objects, you hold down [Shift] and click on each object. Then choose Group from the Arrange menu.

1. Verify that the navigation bar in the main menu scene is displayed.

2. Hold down [Shift] and click on each button to select them all.

3. Choose Group from the Arrange menu.

The buttons are now grouped, as indicated by the set of handles surrounding them. You can ungroup a set of objects by selecting the group and choosing Ungroup from the Arrange menu. Now specify a name and start time for the grouped object. You will set the start time for 0:00.0 to indicate that the navigation bar will appear when the scene begins.

4. Right-click on the navigation bar group and choose Motion.

5. Name the group **navigation bar**

6. Verify that the Start Time value is set to 0:00.0.

7. Click on OK.

Now copy the group and paste it to each subsequent scene.

8. Choose Copy from the Edit menu.

9. Display the treks scene.

10. Choose Paste from the Edit menu.

11. Display the asia scene.
12. Choose Paste from the Edit menu.
13. Continue to paste the group to the remaining scenes.

Now that each scene has the same navigation bar, you need to specify a name for this object for each scene. At the same time, you can verify that the start time is set to 0:00.0 for the navigation bar in each scene.

14. Display the treks scene.
15. Right-click on the navigation bar group and choose Motion.
16. Name the object **navigation bar**
17. Verify that the Start Time value is set to 0:00.0.
18. Click on OK.
19. Display the asia scene.
20. Repeat the steps to name the navigation bar and check its start time for each of the remaining scenes.

The company scene is the last one in the series. Therefore, the Next button will cause the presentation to end. Remove this button to indicate that there are no subsequent scenes.

21. Verify that the company scene is displayed.
22. Verify that the navigation bar group is selected.
23. Choose Ungroup from the Arrange menu.
24. Click on a blank area to deselect the buttons.
25. Click on the Next button.
26. Press [Delete].
27. Save the presentation.

This completes the navigation button settings. Take a moment to see how they will work.

28. Choose Play Presentation from the File menu.
29. Click on the Continue button to display the main menu.
30. Click on Treks to display the treks scene.
31. Click on Company to jump to the company scene.
32. Click on Return to return to the Treks scene (the one that originated the jump to the company scene).

33. Continue testing the navigation buttons. When done, click on Exit to end the presentation.

Adding Text and Linking It to Another Scene

Next you will complete the main menu scene by adding the three options (Treks, Policies, and Our Company). Notice that the three text objects are aligned left, center, and right in the scene. In order to ensure that they are consistent in size, you will create one, resize it, and then copy it to create the other two.

1. Display the main menu scene.
2. Verify that the color chip is set to pink.
3. Use the Font options from the Text menu to change the font to Arial size 24.
4. Use the text tool to draw a box the approximate size shown in figure 13.32.
5. Type **Treks**
6. If necessary, use the alignment buttons in the text edit box to left-align the text (see figure 13.33).
7. Click on a blank area in the scene.
8. Click on *Treks* to select it.
9. Choose Copy from the Edit menu.
10. Choose Paste from the Edit menu.
11. Point to the word *Treks* and drag it down as shown in figure 13.34.
12. Double-click on the second *Treks* to bring up the text edit box.
13. Highlight the word *Treks* and change it to **Policies**
14. Use the right alignment button to move the text to the right.
15. Click on a blank area in the scene.
16. Choose Paste from the Edit menu to paste another *Treks* text object.
17. Move the new *Treks* text to below *Policies* (to the position shown in figure 13.35).

Chapter 13 Developing an Interactive Multimedia Presentation: Part 1

Figure 13.32
Drawing a text box

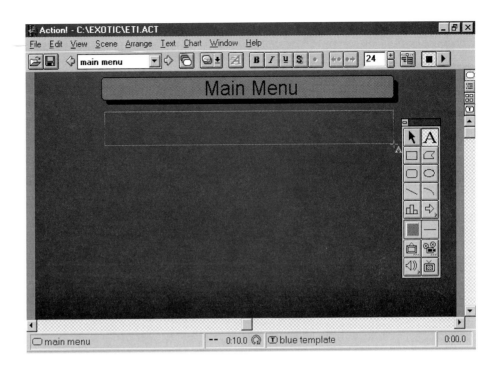

Figure 13.33
Alignment buttons

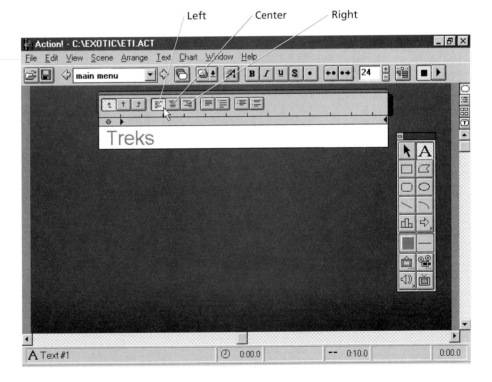

Figure 13.34

Copying the text box

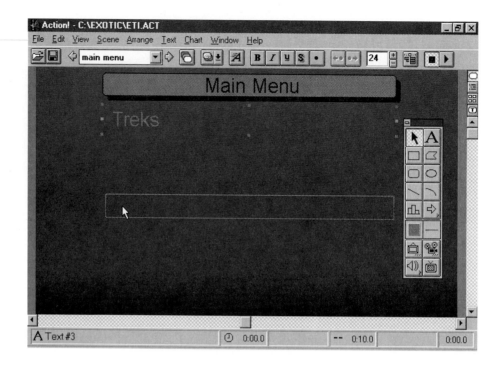

Figure 13.35

Moving the third text box

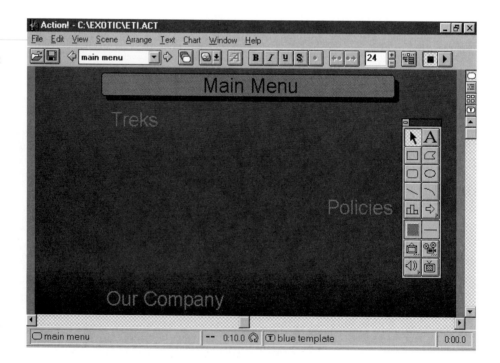

Chapter 13 *Developing an Interactive Multimedia Presentation: Part 1*

18. Repeat the steps needed to change *Treks* to **Our Company** and left-align it.

Your screen should resemble figure 13.35. Now you will link the text items to the appropriate scenes so that a user can use them to navigate through the presentation.

1. Right-click on the word *Treks* and select Interactivity.
2. Change the name to **treks**
3. Activate Highlight and Action.
4. Set the action to Go to Link.
5. Set the destination to treks.
6. Repeat the process for the other two text objects, using the names **policies** and **our company** and setting the appropriate links.

Now check the links by playing the presentation and clicking on each text object.

7. Play the presentation.
8. Click on Continue.
9. Click on Treks.
10. Click on the Return button.
11. Click on Policies and click on Return.
12. Click on Our Company and click on Return.
13. Exit the presentation.
14. Save the presentation.

Next you will add the three text objects (*Asia*, *Africa*, and *Americas*) to the treks scene. Because the layout is the same, you can simply group and copy the text objects from the main menu scene and then edit the names.

15. If necessary, display the main menu scene.
16. Hold down [Shift] and click on Treks, Policies, and (if necessary) Our Company to select each one.
17. Choose Group from the Arrange menu.

18. Choose Copy from the Edit menu.
19. Display the treks scene.
20. Choose Paste from the Edit menu.
21. Choose Ungroup from the Arrange menu.
22. Double-click on the word *Treks* below the heading.
23. Change the word *Treks* to **Asia**
24. On your own, change the word *Policies* to **Africa** and the words *Our Company* to **Americas**
25. Right-click on the word *Asia* and choose Interactivity.

Notice that the settings are for the Treks text object you copied. You need to change them for the Asia object.

26. Change the name to **asia**
27. Change the destination scene to asia.
28. Click on OK.
29. Use the Interactivity panel for the other two text objects to specify a name for the object (**africa**, **americas**) and to set the appropriate links.
30. When done, play the presentation and test the links.

Adding Body Text

Several of the screens have body text that provides useful content for the user. The process for entering text within the body of a scene is essentially the same as adding a single line of text, as you have been doing. Start by adding the body text for the title scene.

1. Display the title scene.
2. Verify that the color chip on the tool palette is pink.
3. Click on the text tool.
4. Choose Font from the Text menu.

Figure 13.36

The text for the title scene

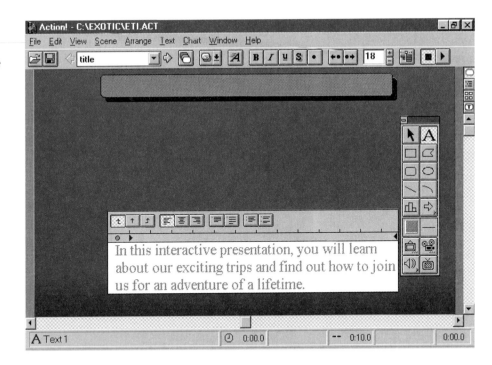

5. Change the font to Times New Roman or a similar font of your choice and the size to 18.

6. Click on OK.

7. Choose Alignment from the Text menu and select Left.

8. Draw a text box near the bottom of the work area and type the text shown in figure 13.36. (*Note:* If you make a mistake, you can edit the text or simply delete it and start again.)

9. Click outside the text edit box to place the text object in the scene.

Now name the text object and check to see that its Start Time value is set to 0:00.0.

10. Right-click on the text and select Motion to display the Object Options dialog box.

11. Change the name from Text #1 to **title scene text**

12. Verify that the Start Time value is set to 0:00.0.

Figure 13.37

The text for the asia scene

Times New Roman bold 18

Times New Roman 18

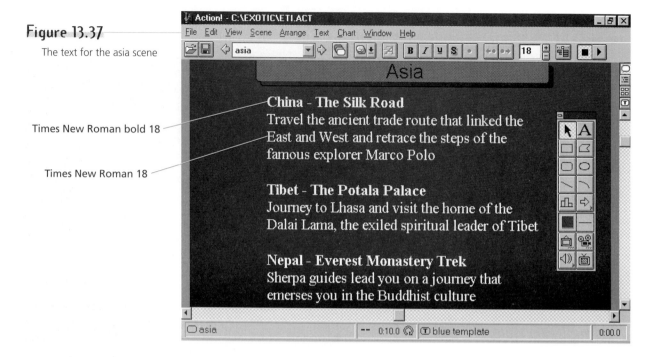

13. Click on OK to return to the scene.

Figure 13.37 shows the text for the asia scene including the font and type size. Notice that the headings (China - The Silk Road, etc.) are in bold. Figure 13.38 shows the four other scenes that have text. You will complete this part of the presentation by entering the text for these five scenes.

14. Complete the text for the asia, africa, americas, policies, and our company scenes. Change the name of each text object as appropriate (**asia scene text**, and so on) and verify that the Start Time value is set to 0:00.0.

15. Save the presentation.

16. Play the presentation.

17. Exit the presentation.

Chapter 13 Developing an Interactive Multimedia Presentation: Part 1

Figure 13.38

The text for the final four scenes

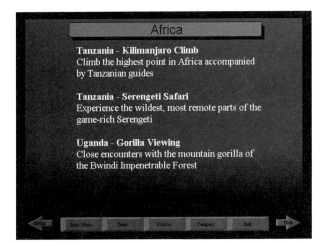
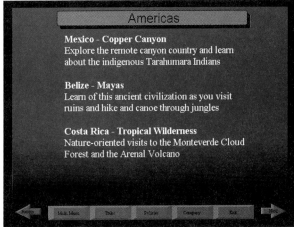

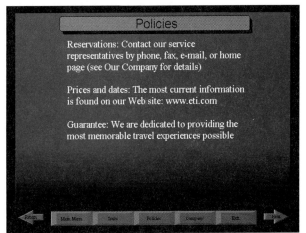

Getting Help While Using Action!

Action! provides a Help feature to assist you in using the program. Although there are several parts to the Help feature, this section focuses on the common methods used to get help. Action! provides both context-sensitive help and a help topics list similar to other applications.

The context-sensitive help allows you to display help screens relevant to the process you are currently working with or a tool that you want to know more about. For example, if you wanted to know what could be done using a particular tool in the tool palette, you could select the help pointer from the Help menu and click on the tool. Try this now.

1. Choose Pointer from the Help menu.

The mouse pointer changes to a question mark. Placing the question mark over a tool and clicking the mouse button will display the help screen for that tool.

2. Click on the selection tool in the tool palette.

The Selection tool help screen appears, indicating what the tool is used for.

3. Click on Resizing objects.

The help screen explains how the selection tool can be used to resize an object.

4. Choose Exit from the File menu in the Help window to exit Help.

Now use the Topics command in the Help menu to search for a topic and get help on it.

5. Choose Topics from the Help menu.

The Action! 3.0 Help window appears. The window has five parts:

- The **title bar** indicates that this is the Action! help window.

- The **menu bar** is similar to the application menu bar and has four menus: File, Edit, Bookmark, and Help.

- The **tool bar** has the following four options:
    - **Index** provides an alphabetical list of all topics and allows searching on a topic.
    - **Go Back** is used to return to a previously viewed topic.
    - **History** displays a list of all of the topics viewed.
    - **Search** is used to search for a topic.

- The **Contents section** displays a list of menus and main features that can be selected.

- The **Main section** displays, in this case, how to use the Help feature.

1. Click on Index.

A window appears with an alphabetical Index list and a number indicating how many topics are associated with each item in the list. You can scroll down the list or type a word to search for.

2. Type **play**

Notice that as you type, the topics list scrolls to display the closest match for your search word. Play command is highlighted with five topics.

3. Click OK to select Play command.

The five topics for Play command are listed. These are the different ways you can play a scene or presentation.

4. With control panel highlighted, click on Go To.

The Control panel help screen appears.

5. Click on the History button.

A list of the help screens that have been displayed appears. You could choose any of them.

6. Click on Cancel to remove the History window from view.

7. Choose Exit from the File menu to close the Help window.

This completes the section on using the Help feature.

Printing a Presentation

In this section you learn about the printing options available in Action! and how to print a presentation. Before printing you should consider the type of printer available. If you do not have a color printer, you may want to omit the background color to achieve a more readable printout.

1. Choose Print from the File menu.

The Print dialog box appears, specifying the default printer and allowing you to indicate what to print, how many copies, the print range, and print time. The dialog box also allows you to omit the background, print to a file, and collate copies. The only change you will make is to specify that four scenes be printed per page.

2. Click on the down arrow in the Print What box.

A drop-down list appears, allowing you to specify the number of scenes to print per page or to print other views of the presentation.

3. Click on (four per page).

4. Select Omit Background if you do not have a color printer.

5. Make sure the printer is ready.

6. Click on OK.

The scenes are printed four to a page.

In this chapter you have learned how to develop multimedia titles using Action! 3.0. You know how to create backgrounds and scenes and how to set properties for various objects. You are also able to create buttons for hyperlinking and specify motions and effects for objects. With these basic skills, you are prepared for the more advanced features of Action! presented in the next chapter.

To extend what you've learned, log on to the Internet at

http://www.thomson.com/wadsworth/shuman

You will find a wide variety of resources and activities related to this chapter.

review questions

1. **T F** Action! uses a time-based process for displaying scenes.

2. **T F** Tool palette is another name for the tool bar.

3. Every element, such as a button or picture, that is placed in a scene is called a(n) _____.

4. **T F** The Enter setting for a text object indicates when the text will appear in a scene.

5. The control panel has two buttons that determine which objects are in view when the motion in a scene is stopped: the _____ view and the _____ view.

6. **T F** Hyperlinking is done with the Interactivity options.

7. **T F** The Skip To actions are used with the Return action.

8. **T F** Using the new scene button creates a new scene with the current scene's template.

9. When you search the help index using the word *objects,* how many topics are displayed?

10. When you use the help pointer and click on the text tool in the tool palette, what related topic is mentioned?

projects

1. Create an interactive résumé for yourself that includes the following features:
   - A template you create
   - At least six scenes
   - A shadowed text box
   - A navigation bar
   - At least two different fonts and at least two different type sizes
   - At least three different object effects and three different transition effects
   - Use a sparkle or shimmer effect for a text object

   Save the presentation as **resume** with your initials (for example, *resumejs*).

2. Create an interactive informational presentation about your school. The content should have data about programs and services and how to get more information. Include the same features as listed in project 1.

   Save the presentation as **school** with your initials (for example, *schooljs*).

3. Complete the Guided Tour that comes with the Action! 3.0 program.

Developing an Interactive Presentation: Part 2

AFTER COMPLETING THIS CHAPTER YOU WILL BE ABLE TO:

- Use the advance features of Action! 3.0
- Create hotword masks for hyperlinking objects
- Add graphics, sounds, and movies to a presentation
- Add motion to graphics
- Create path animations
- Work with the Timeline feature

N this chapter you learn how to incorporate more-advanced features of Action!, including creating embedded text with hyperlinking, adding graphics with hyperlinks, adding sound, creating animations, and incorporating video.

Hyperlinking Using Embedded Text

In chapter 13 you learned how to specify a link with a text object such as the options (Treks, Policies, Our Company) displayed in the main menu scene. In this section you learn how to specify a word or phrase within a paragraph as a *hotword*, that is, a word that causes some type of action (go to another scene, play a sound, and so on). Figure 14.1 shows the policies scene. Notice that the first paragraph says "see Our Company for details." You will make the phrase *Our Company* a hotword so that when the user clicks on it, the our company scene will be displayed.

1. Open the Exotic Treks presentation.

2. Display the policies scene.

You cannot select individual words to link, only entire text objects. You can, however, place another object, such as a transparent rectangle, over the desired text and specify the desired interactivity for the object. Then when

Figure 14.1

The policies scene

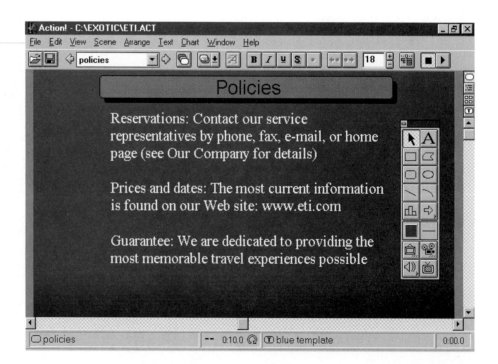

Chapter 14 *Developing an Interactive Multimedia Presentation: Part 2* **313**

the user clicks on what appears to be the hotwords, the transparent object is activated. It is important to indicate to the user that the words are "hot" by having them appear different than the other words. This can be done by using a text style (bold, italic, underline), a different font, color, or type size, or any other way to make the hotword(s) stand out. In this case, you will apply an italic type style to the words.

3. Double-click on the word *Company* to open the text edit box.

4. Drag the pointer across the words *Our Company* to select them.

5. Choose Italic from the Style options in the Text menu or click on the italics button on the tool bar.

6. Click on a blank area to remove the text edit box.

Now create a transparent object that will cover the text.

7. Use the rectangle tool to draw a box that covers the words *Our Company* (see figure 14.2).

8. Right-click on the box and select Visual.

9. Change the name of the object to **hotword mask**

10. Change the Ink Mode option to Transparent.

11. Verify that the Fill Pattern option is set to Standard.

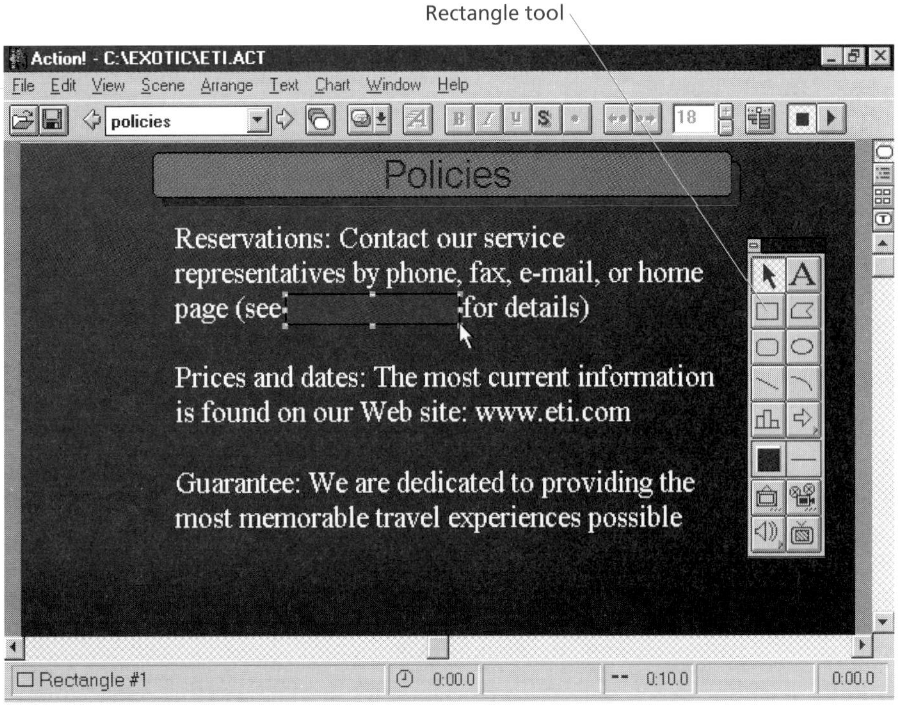

Figure 14.2

Drawing an object that covers the words *Our Company*

12. Click on the color chip next to the word *Standard* in the Fill Pattern.
13. Choose N for none.

Specifying none for the Fill Pattern removes the color within the object. Specifying Transparent for the Ink Mode removes the border for the object. Now specify a link for this object.

14. Click on the Interactivity tab.
15. Activate the Highlight and Action options.
16. Set the action to Go to Link.
17. Set the destination scene to our company.
18. Click on OK.

Now test the hotword link.

19. Play the presentation.
20. Display the policies scene.
21. Click on Our Company.
22. Exit the presentation.

Adding Graphics with Hyperlinks

In this section you learn how to add graphics and specify properties such as sounds and hyperlinks. Figure 14.3 shows the main menu with three graphics added. The process is to choose the import picture tool from the tool palette, specify which graphic to import, and move the graphic to the desired location. Once a graphic appears, it is considered an object and its properties (size, links, transitions) can be set. Start by importing the first graphic for the main menu scene.

1. Display the main menu scene.
2. Click on the import picture button in the tool palette.

The Import Picture dialog box appears, allowing you to specify which picture to import. Depending on your computer setup, you need to specify the correct directory or folder that contains the *trek.bmp* file.

3. Complete the dialog box to specify the *trek.bmp* file to import.
4. Click on OK.
5. Move the graphic to the location shown in figure 14.4.

Figure 14.3

The main menu scene with three graphics added

Import picture tool

Figure 14.4

Placing the imported picture

6. On your own, import and place the other two graphics: *policies.bmp* and *ourcomp.bmp*.

Now you will link these graphics to their respective scenes.

7. Right-click on the *trek.bmp* graphic and choose Interactivity.
8. Change the object name to **treks.bmp**
9. Activate the Highlight and Action options.
10. Set the action to Go to Link and the destination scene to treks.
11. Click on OK.
12. On your own, specify the appropriate names and links for the other two objects.
13. Play the presentation and test the graphic links.
14. Save the presentation.

Adding Motion to Graphics

The same types of motion that can be applied to text (wipes, pulls, grows, and so on) can be applied to graphics. In this case, you will specify a motion for the Exotic Trips logo in the title scene.

1. Display the title scene.
2. Import the picture *logo.bmp*.
3. Resize and position the graphic as shown in figure 14.5.
4. Display the Motion panel for this object.
5. Name the object **logo.bmp**

You will have the logo enter the scene after the title appears. The title begins to appear when the scene appears and takes 1 second to be pulled in. So you will set the start time for the logo to be 1 second.

6. Set the Start Time value to 1 second.
7. Set the Enter time (the amount of time it takes the object to enter the scene) to 2 seconds.
8. Set the Enter effect to Grow From Center.
9. Verify that the Play Time is set to To End of Scene.

Figure 14.5
Importing the logo picture

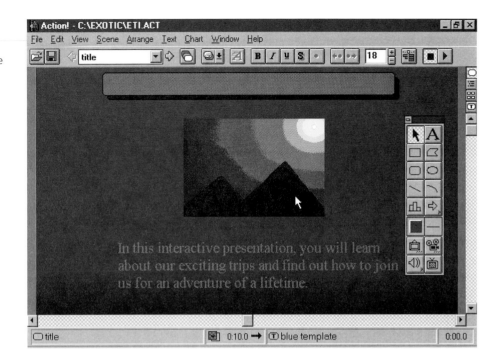

10. Exit the Object Options dialog box.
11. Play the presentation and then save it.

Adding Sound to a Presentation

Action! allows you to specify a sound that can be played anytime during a presentation. The sound can be played as a scene starts, holds, or ends, or it can be associated with an object such as a button, text, or graphic. The process is to select the object that will cause the sound to be played and use the Sound panel to specify which sound and when it should be played. If the sound is not to be associated with an object, the import sound tool in the tool palette can be used to specify the sound file and how long it will play during a specific scene. In this case, you will specify that a sound be played when the logo is displayed.

1. Display the title scene and, if necessary, use the control panel to advance the scene to display the logo.
2. Right-click on the logo and choose Sound.
3. Name the object **intro.wav**

Figure 14.6

The Sound panel

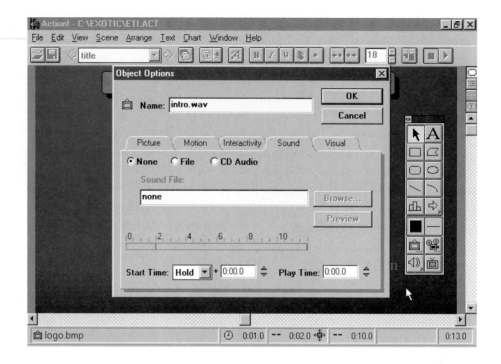

Figure 14.6 shows the Sound panel. Study the Sound panel as you read the following description.

Sound source You can specify that a sound be played from a computer file (either a wave audio file or a MIDI sound file) or you can specify that a sound be played from a CD.

Sound File This option is used to indicate which file to play. You could use the Browse button to locate the file.

Playback options Here you can indicate when the sound starts to play in relation to the object. That is, if you want the sound to play as the object enters the scene, you would specify 0:00.0 as the start time. If you wanted to have the sound play 2 seconds after the object enters the scene, you would specify 0:02.0 as the start time. You can also indicate the play time. If you want the sound to play longer than its actual length, it will loop until the specified time is up.

You will need to specify the location of the sound file called *intro.wav* and verify that the start time is 0:00.0.

   4. Click on the File option button.

Figure 14.7

The completed Sound panel

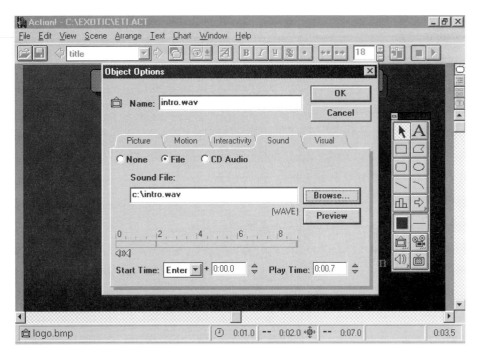

5. Click on the Browse button and complete the Import Sound dialog box to indicate the location of the *intro.wav* file. When done, Click on OK to close the dialog box and return to the Sound panel.

6. Set the Start Time value to Enter+0:00.0.

7. Verify that your screen resembles figure 14.7. (*Note:* The location of the file may be different for you.)

8. Click on OK to close the Object Options dialog box.

9. Verify that your computer's sound system is ready to play a sound.

10. Play the presentation.

11. Exit and save the presentation.

Creating Animations

Action! allows you to create simple 2-D path animations. In this section you learn how to create a path animation that will cause an object to move across the screen in a path you specify. Figure 14.8 shows a new scene that displays a map. When the user clicks on Lhasa in the asia scene, an animation plays

Figure 14.8

A new scene with a map

that shows the path of the plane for each day of travel. When the user clicks on Potala Palace in the asia scene, an arrow pointing to the final destination and a picture representing the destination appear.

Action! provides several ways to animate objects, including ones you have already used, such as Pull Left and Grow From Center. *Path animations* are the most flexible, however, because they allow you to move an object in any direction and change the direction as often as desired. There are two important things to keep in mind when creating a path animation: the number of times the object is moved and the amount of time specified for each movement.

A *nonlinear animation* is made up of a series of small movements, called *segments*. The more times the object is moved, the smoother the animation appears. And the more time each segment takes, the longer the entire animation runs. Study figure 14.9 which shows the animation you will create. The arrow starts at the left and moves to the right. Each arrow represents a segment, and each segment has a specified time. In this case, each segment is 0.2 second. You will create this animation using the Path Editor. Start by creating a new scene and importing the map as a background for the scene. Placing the map on the background helps ensure its visibility when the animation is being created.

Chapter 14 *Developing an Interactive Multimedia Presentation: Part 2* **321**

Figure 14.9
A nonlinear path animation

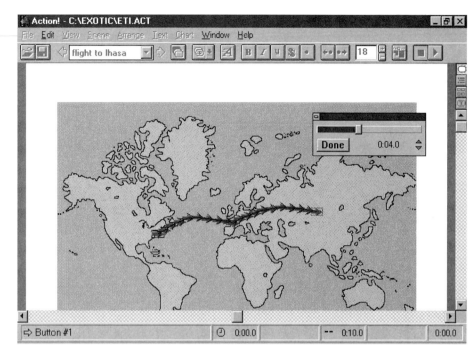

1. Display the our company scene.
2. Choose New from the Scene menu.
3. Choose Options from the Scene menu.
4. Name the scene **flight to lhasa**
5. Click on the Time tab.
6. If necessary, set the Scene Length value to 10 seconds and select the As Long as Objects option.
7. Click on the Background tab in the Scene Options dialog box.
8. Click on Picture.
9. Click on Browse.
10. Use the Import Picture dialog box to import the *map.bmp* picture.
11. Click on OK in the Scene Options dialog box to return to the scene.

Now you will use an arrow button for the object to animate.

12. If necessary, change the color chip in the tool palette to pink.
13. Click on the button tool and select the right-pointing 3-D arrow.

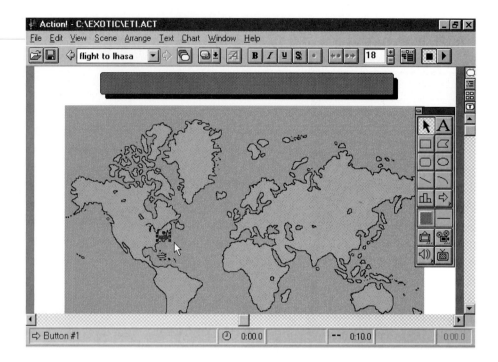

Figure 14.10
Drawing an arrow on the map

14. Draw an arrow approximately the size of the one shown in figure 14.10.

15. Right-click on the arrow and choose Motion.

16. Change the name of the object to **animated arrow**

17. Click on the From button in the Enter box.

Notice that the list of options includes Start/End and Path Editor. Start/End can be used to create straight-line animation. The Path Editor is used to create a nonlinear animation.

18. Click on Path Editor.

The screen appears with the Path Editor box. The Path Editor is used to set the timing for each movement in the animation. That is, you must specify a duration time for each segment of the path. In this case, each time you move the arrow you will increase the duration by 0.2 second. Start by setting the original location for 0.2 second.

19. Click on the up arrow in the Path Editor to set the duration to 0.2 second (see figure 14.11).

20. Drag the arrow to the next location as shown in figure 14.12.

Chapter 14 *Developing an Interactive Multimedia Presentation: Part 2* **323**

Figure 14.11

Setting the time for the animation segment to 0.2 second

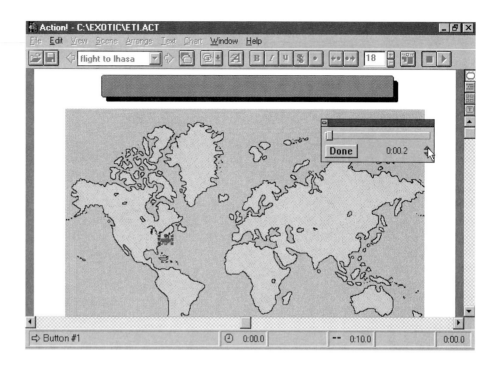

Figure 14.12

Dragging the arrow to create the next animation segment

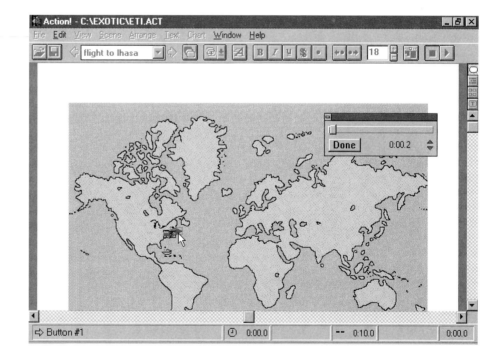

21. Add 0.2 second to the Path Editor.

22. Continue to drag the arrow and add 0.2 second to the Path Editor as you draw an approximation of figure 14.9. When you have completed the animation, be sure to add 0.2 second for the last segment and then click on Done.

Note: Proceed slowly and make sure you have the same number of segments as shown in Figure 14.9. If you make a mistake, you can edit the animation, but it is usually easier to begin over by clicking on the Done button in the Path Editor and then deleting the object and redrawing it.

After you click on Done, the Motion panel of the Object Options dialog box appears. Check the enter time, which indicates how long the animation will take. You need to set a pause for the scene at the exact time the animation ends.

23. Note the enter time in the Motion panel.

24. Click on OK.

25. Choose Add Pause from the Scene menu.

26. Verify that the pause is set to the time needed for the animation.

27. Click on OK.

Now use the control panel to play the animation.

28. Choose Control Panel from the Window menu to display it.

29. Click on the point-in-time button to show the objects as they appear in the scene at any specific time.

30. Click on the rewind button in the control panel.

31. Click on the play button.

32. Save the presentation.

Next you will add two text objects to the scene. One will say *Day 1: New York to Paris* and the other will say *Day 2: Paris to Lhasa*. Each one will be timed to appear during the appropriate leg of the journey.

1. Draw a text box and enter the words *Day 1: New York to Paris* (see figure 14.13).

2. Select the words and change the type size to 14 and bold.

Now you will need to specify when the text object will appear and disappear from the screen. It should appear as the animation begins and disappear

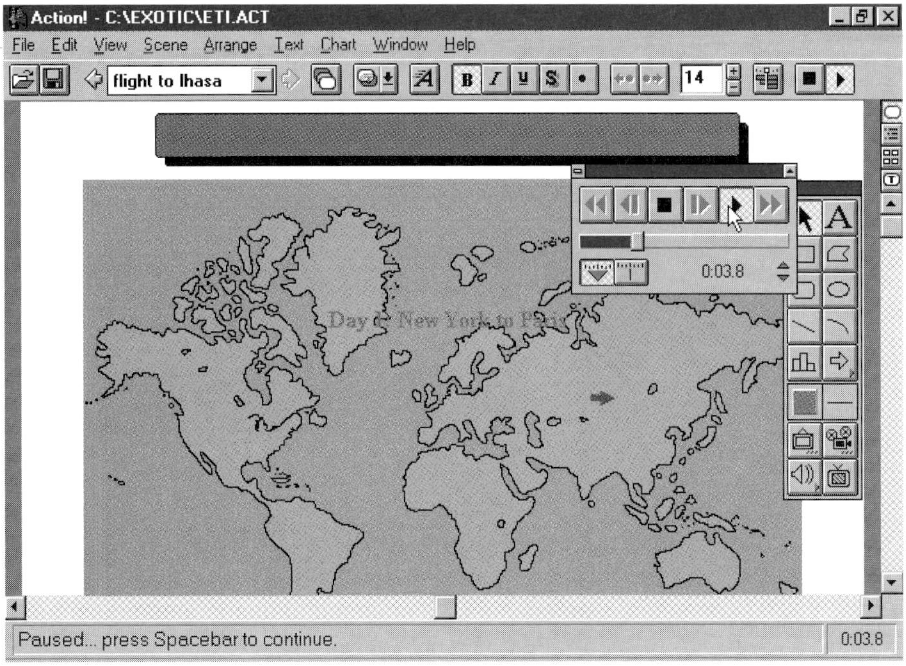

Chapter 14 *Developing an Interactive Multimedia Presentation: Part 2* **325**

Figure 14.13
Adding text to the scene

when the arrow reaches Paris. You can use the timing buttons on the control panel to determine when the arrow reaches Paris.

3. Rewind the scene.

4. Click on the timing buttons until the arrow points to Paris (see figure 14.14).

5. Note the elapsed time (figure 14.14 shows 1.7 seconds, but yours may be different).

6. Click on the span-time button in the control panel to display the text (see figure 14.14).

7. Right-click on the text object and choose Motion.

8. Change the name of the text object to **day 1**

9. Set the start time to 0:00.0 and the play time and hold time to the time it took for the arrow to reach Paris.

10. Click on OK.

Now you can copy the text object, move the new text object to below the first one, and edit it to read *Day 2: Paris to Lhasa* (see figure 14.15).

11. Click on the text object to select it.

Figure 14.14

Using the timing buttons to move the arrow

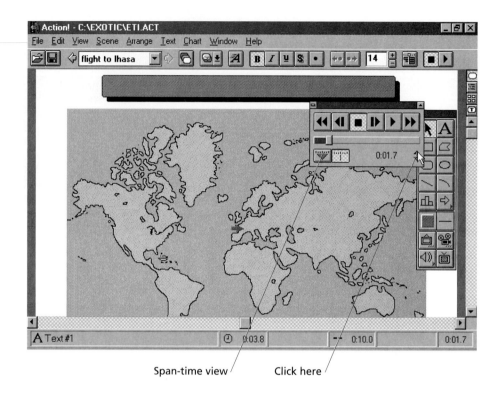

Span-time view   Click here

Figure 14.15

The day 2 text object

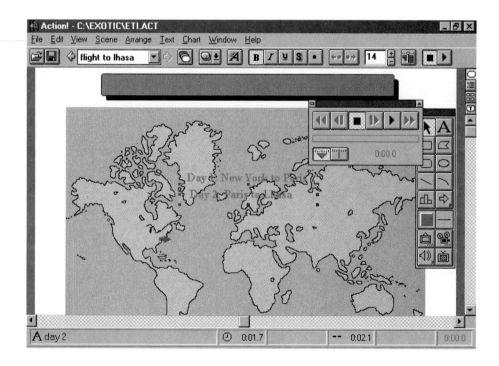

12. Choose Copy from the Edit menu.
13. Choose Paste from the Edit menu.
14. Drag the copied object below the first object.
15. Double-click on the copied object and change the text to **Day 2: Paris to Lhasa**
16. Click on a blank area in the scene.

Now you need to set the timing for the day 2 object. You want it to start when the day 1 text disappears, that is, whenever your animation reaches Paris, and you want it to hold to the end of the animation.

1. Click on the point-in-time button in the control panel and rewind the scene.
2. Right-click on the day 2 object and choose Motion.
3. Name the text object **day 2**
4. Set the start time to the time the animation took to reach Paris.
5. Set the hold time to the difference between the start time and the end of the animation. For example, if the start time is 1.7 seconds and the end of the animation is 3.8 seconds, set the hold time for 2.1 seconds.
6. Click on OK.
7. If necessary, use the control panel to rewind the scene. (*Note:* You may need to click on the point-in-time button in the control panel.)
8. Use the control panel to play the scene.
9. Rewind the scene and save the presentation.

Before leaving this scene, you will enter a text heading and copy the navigation buttons from the main menu scene.

1. Use the text tool to enter the text heading **Flight to Lhasa** using a black color and Arial size 24. Center the text.
2. Display the main menu scene.
3. Scroll down to view the navigation buttons.
4. Click on the navigation buttons to select them. (*Note:* They should still be grouped.)

5. Choose Copy from the Edit menu.
6. Display the flight to lhasa scene.
7. Choose Paste from the Edit menu.
8. Right-click on the navigation bar and choose Motion.
9. Change the name to **navigation bar**
10. Verify that the Start Time value is set to 0:00.0.
11. Click on OK.
12. Save the presentation.

Next you need to link the asia scene to this one.

13. Display the asia scene.
14. Change the word *Lhasa* to italic type.
15. If necessary, change the color chip to pink.
16. Draw a square-cornered rectangle over the word *Lhasa*.
17. Right-click on the rectangle object and choose Visual.
18. Change the name of the object to **lhasa hotword mask**
19. Change the Ink Mode option to Transparent.
20. Click on the color chip near the Fill Pattern option.
21. Click on N to specify none for the fill color.
22. Click on the Interactivity tab.
23. Activate Highlight and Action.
24. Set the action to Go to Link and the destination scene to flight to lhasa.
25. Click on OK.
26. Play the presentation, jump to the asia scene, and choose the Lhasa hotword.
27. After viewing the animation, exit the presentation.
28. Save the presentation.

The next scene you add shows how the grow effects can be used to animate an object. In this case, you will create a new scene that the user can jump to by clicking on the words *Potala Palace* in the asia scene. The new scene will have an arrow grow up from the bottom of the scene and point to the place on the map where the Potala Palace is located. Then a picture of the palace will grow out from the map. Start by adding a new scene.

1. If necessary, display the flight to lhasa scene.
2. Choose New from the Scene menu.
3. Choose Background from the Scene menu.
4. Change the scene name to **potala palace**
5. Click on Picture.
6. Click on Same as.
7. Click on flight to lhasa and click on OK.
8. Click on the Time tab and, if necessary, set the scene length to 10 seconds.
9. If necessary, select the As Long as Objects option.
10. Click on OK.
11. Set a pause for the scene at 9 seconds.

Continue by pasting the navigation bar into this scene and changing the interactivity for the Next button so it will send the user to the company scene. Because this is the last scene, if you leave the Next button set to go to the next scene, Action! will exit the presentation.

12. Paste the navigation bar into this scene.
13. Ungroup the navigation bar and set the interactivity for the Next button to go to the company scene.

Now draw the arrow as shown in figure 14.16 and set the effect and timing for the arrow.

14. Verify that the color chip is pink.
15. Choose the 3-D up arrow from the button palette in the tool palette.
16. Draw the arrow shown in figure 14.16.
17. Right-click on the arrow and choose Motion.
18. Change the name of the button to **lhasa arrow**
19. Set the enter time to 3 seconds. This will cause the arrow to take 3 seconds to grow.

Figure 14.16

Drawing an arrow

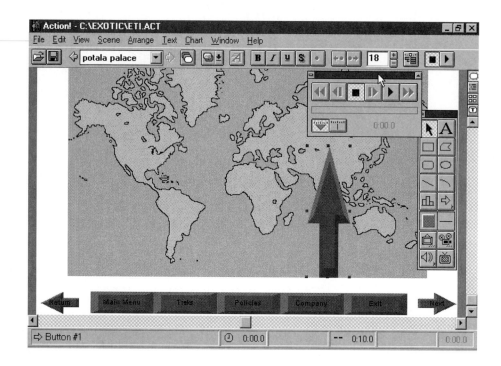

20. Set the hold time to 1 second.
21. Click on the Effect button in the Enter box.
22. Choose Grow and specify Grow From Bottom.
23. Click on OK.

Now import a picture, place it on the map where the arrow is pointing, and have it grow out from the center when the arrow completes its motion.

24. Import the picture named *potala.bmp.*
25. Place the center of the picture over the tip of the arrow.
26. Right-click on the picture and choose Motion.
27. Change the name to **potala.bmp**
28. Set the start time to 3 seconds (the time the arrow has completed its motion).
29. Set the enter time to 4 seconds (the time it will take to grow the picture).
30. Set the enter effect to grow from the center.
31. Click on OK.
32. Use the control panel to play the scene.

33. Save the presentation.

Finally, link the asia scene to this one and enter a heading for this scene.

34. Display the asia scene.

35. Create a hotward link for the words *Potala Palace.* Be sure to italicize the words, name the rectangle object (**potala hotword mask**), set the fill pattern to none, and the ink mode to transparent.

36. Link the hotword to the potala palace scene.

37. Display the potala palace scene and enter the text **Potala Palace** in the heading box. Set the color to black and the font to Arial size 24, and center the heading.

38. Set the start time for the text to 0:00.0 and the name to **potala palace**

39. Play the presentation.

40. Jump to the asia scene.

41. Click on Potala.

42. After viewing the animation, exit the presentation.

43. Save the presentation.

Playing Video and Animation Clips

Some of the most compelling elements in a multimedia title are movie and animation clips. Action! allows you to include these elements in any scene and control when and for how long they are played. You can even specify a portion of the clip to be played. Action! supports the following movie and animation formats.

Format	Filename Extension	Type
QuickTime	.mov	Digital video
Microsoft Video for Windows	.avi	Digital video
Autodesk Animator	.fli and .flc	Animation
Macromedia Director Player	.mmm	Animation

Figure 14.17

Importing a movie

Note the following specifics about video and animation clips:

- Movies may have their own color palette, and Action! may give you the option of importing the palette when importing the movie.
- When importing a Director movie, any interactivity in the movie is lost.
- Do not import two separate movies if they have different palettes.
- Do not play more than one movie at a time.
- Movie files can be very large and often require a large amount of computer memory to play properly.

You will import a movie that will be placed in the treks scene and be started by clicking on a movie icon. The process for importing a movie is similar to importing a picture.

1. Display the treks scene.
2. Click on the import movie tool in the tool palette.
3. When the Import Movie dialog box appears, choose the *asia.avi* movie.
4. Click on OK.
5. Position the movie as shown in figure 14.17.
6. Right-click on the movie and choose Motion.

Figure 14.18

Importing a picture to be used as a play/pause button

7. Change the name to **asia.avi**
8. Set the start time to 0.2 second. This will allow you to set a pause at 0.1 second, and then the movie icon can be used to start and stop the movie.
9. Click on OK.
10. Set a scene pause at 0.1 second.

Now import a picture that will be used as the play/pause movie icon.

11. If necessary, choose the span-time button in the control panel.
12. Import the picture named *movicon.tif*.
13. Position the picture as shown in figure 14.18.
14. Right-click on the movie icon and choose Interactivity.
15. Change the name to **movie icon**
16. Activate the Highlight and Action options.
17. Set the action to Pause/Continue.
18. Click on OK.
19. Select the point-in-time button in the control panel and play the scene.

20. When the scene pauses, click on the movie icon to play the movie.

21. Use the timing button in the control panel to determine when the movie ends.

22. Choose Options from the Scene menu and click on the Time tab.

23. Set the scene length in the Time panel to the same time as the moment the movie ends. This will keep the movie from replaying.

24. Click on OK.

25. Play the movie and click on the movie icon several times.

26. Save the presentation.

It is important to understand that the movie icon is not linked directly with the movie. Rather it is used to pause and continue the playing of the scene.

Editing a Movie

Action! allows you to edit an *.avi* file in the following ways:

- Selecting a part of the movie clip to play
- Changing the sound volume
- Specifying the playback speed

These tasks are done using the Digital Video Editor. Figure 14.19 shows the Digital Video Editor and identifies the parts of the editor. Notice that the displayed movie plays from 0 to 98. Changing the 0 to a higher number starts the movie later, and changing the 98 to a lower number stops the movie sooner. Changing these settings therefore allows you to select any segment of the video clip to play. Remember, this works only with *.avi* files.

Working with the Timeline

Action! provides a way to graphically represent the timing of each object within a scene, that is, when an object appears, how long it stays in the scene, and when it disappears. This is useful in synchronizing the action of several objects within a scene. By displaying a scene's timeline, you can easily edit the timing of each object. Figure 14.20 shows the timeline for the main menu scene. Near the top of the Timeline window is the time scale with

Chapter 14 Developing an Interactive Multimedia Presentation: Part 2   **335**

Figure 14.19

Editing an *.avi* file

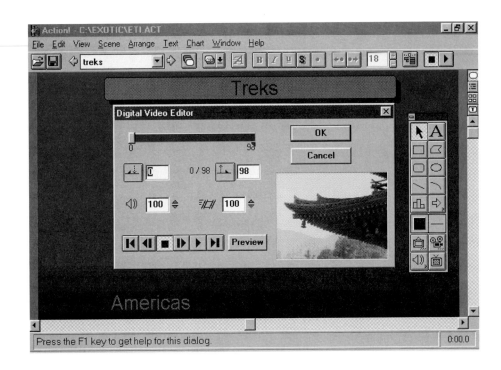

Figure 14.20

The timeline for the main menu scene

Playback head

Zoom tools

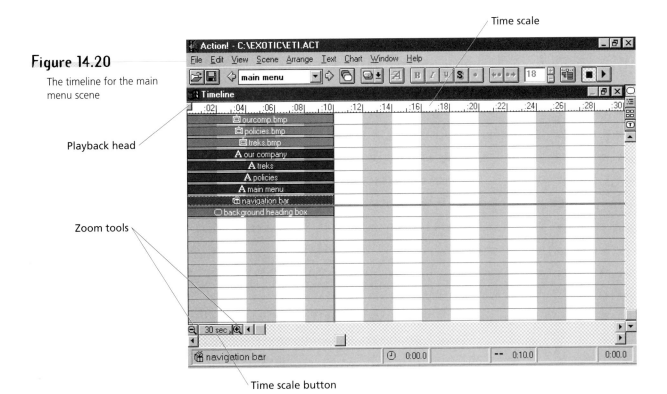

Time scale

Time scale button

increments in seconds. Each row of the timeline displays an object and includes the following:

- The object's name

- An icon representing the type of object (the icon usually is the same as the one in the tool palette used to represent the type of object, for example, *A* representing text)

- A colored bar that shows:
    - When the object appears in and disappears from the scene
    - The type of object: text=blue, visual objects (drawn and imported graphics, movies)=gold, buttons=red, sounds=turquoise, grouped objects=magenta
    - The enter, hold, and exit phase of an object

Other parts of the timeline are labeled in figure 14.20 and described below.

Playback head This button moves across the time scale as the scene is played. You can move it manually to see what is happening at any one time in the scene.

Pause control This button indicates at what time in the scene the play will be paused (not shown in figure 14.20).

Time scale button This button indicates the length of time that will be displayed on the time scale. You can set the time scale to 10 or 30 seconds and 1, 5, 10, or 30 minutes.

Zoom tools These tools are used to enlarge or reduce the timeline to see more detail.

You will use the timeline to edit the main menu scene and cause the three pictures to appear at different times. The timeline button on the tool bar is used to display the timeline (*Note:* It is also used to remove the Timeline window from display on the screen.)

1. Display the main menu scene.
2. Click on the timeline button on the tool bar.

Now resize the Timeline window so that the scene and the window are both displayed.

Figure 14.21

Positioning the scene and the timeline

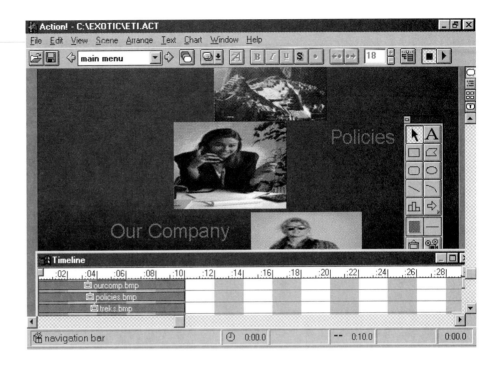

3. If necessary, click on the minimize button on the Timeline window.

4. Scroll the scene and move the Timeline window to approximate figure 14.21.

The timeline shows that there are three graphics, four text objects, the navigation bar, and a background text box (the one used for the headings). Take a moment to play the scene and watch the playback head move across the time scale.

5. Click on the play button in the upper-right corner of the tool bar (see figure 14.22).

6. Watch as the playback head moves across the time scale.

Each object enters the scene at the start of the scene and holds until the end. You will change the three graphic objects so that they enter at 2-second intervals. You can change the timeline of an object by dragging its colored bar. Start by changing the start time for the trek.bmp object to 2 seconds.

7. Click on the stop button on the tool bar (see figure 14.22).

8. Rewind the scene using the control panel.

9. Point to the far-left side of the treks.bmp colored bar (see figure 14.22).

10. When the pointer changes to a double arrow, hold down the mouse button and drag the bar to the 2-second mark on the time scale. (*Note:* This may take several tries.)

11. Repeat the process for policies.bmp and ourcomp.bmp, moving them to the 4-second and 6-second marks, respectively (see figure 14.23).

Notice that the pictures disappear from the scene because they are not specified to be at the start of the scene. Now play the scene and watch the playback head as it moves to the 2-, 4-, and 6-second marks and the pictures appear.

12. Click on the play button on the tool bar.

13. Stop the presentation.

14. Save the presentation.

15. Click on the timeline button on the tool bar to close the Timeline window.

16. Choose Play Presentation from the File menu.

17. Click on Continue and watch as the pictures appear every 2 seconds.

18. Exit the presentation.

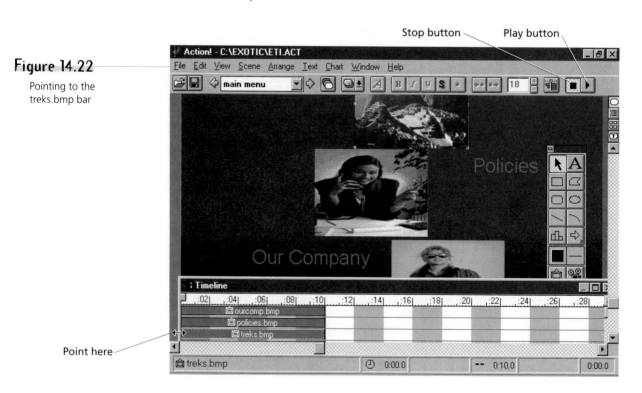

Figure 14.22
Pointing to the treks.bmp bar

Chapter 14 *Developing an Interactive Multimedia Presentation: Part 2* **339**

Figure 14.23

The completed changes to the Timeline window

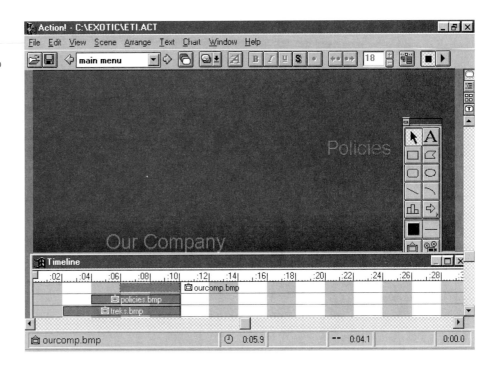

To extend what you've learned, log on to the Internet at
http://www.thomson.com/wadsworth/shuman
You will find a wide variety of resources and activities related to this chapter.

review questions

1. **T F** Action! allows you to create hotwords that are used for hyperlinking.

2. **T F** The same types of motion that can be applied to text (wipes, pulls, grows) can be applied to graphics.

3. **T F** Action! allows you to create only linear animations.

4. **T F** File formats supported by Action! include QuickTime movies and Director animations.

5. **T F** Action!'s Digital Video Editor allows you to select parts of a movie to play.

6. **T F** The timeline can be used to synchronize the movement of various objects in a scene.

7. **T F** Action! allows you to create hyperlinks using text, but not using graphics.

8. **T F** A sound can be played as a scene starts, holds, or ends.

9. **T F** The more segments in an animation, the smoother the animation.

10. When creating an animation, you need to keep in mind two things:

   a. _____

   b. _____

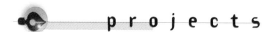

projects

1. Edit the Exotic Treks presentation as follows:
   - When the user clicks on Exit in the navigation bar, a scene appears, asking the user to confirm (by clicking on either a Yes or a No button) the wish to exit the presentation. A yes click exits the presentation, and a no click returns the user to the original scene.
   - Insert pictures for the Africa and Americas options on the main menu.

2. Edit your résumé developed in project 1 of chapter 13 to include at least:
   - 3 pictures
   - 1 path animation
   - 1 sound clip
   - 1 hotword mask

   Save the presentation as **resume2**

3. Edit the school informational title developed in project 2 of the previous chapter to include at least:
   - 3 pictures
   - 1 path animation
   - 1 sound clip
   - 1 hotword mask

   Save the presentation as **school2**

4. Use Action! to develop a presentation of your own design. This could be related to work, a hobby, a class, or any other subject you choose. Use at least 10 scenes and two backgrounds. Include all of the features (video is optional) you have learned in this and the previous chapter.

GLOSSARY

alpha testing The first formal testing of a multimedia title. It is usually conducted in-house and is not restricted to the development team. The idea is to "try to make it crash," and every conceivable action (point and click) and navigation path is explored. See also *beta testing*.

analog wave pattern A representation of the recurring pattern created by a sound.

animation The perception of motion created by the rapid display of still images.

antialiasing The process of smoothing the edges of a bitmap by blending the colors on the edge of the image with the adjacent colors, thereby improving the quality of the image display.

appearance That part of the user interface related to how each object is arranged on the screen.

applet An application that runs on the World Wide Web, such as a program that tracks stock prices and periodically displays them on a Web page.

asymmetrical A method for achieving balance in screen design through the arrangement of dissimilar elements.

audio card A hardware device installed in a computer that allows digital sounds to be played through a speaker. The type and features of the audio card determine the quality of the sound.

authoring program Primary development tools for creating multimedia titles that allow you to combine elements (text, graphics, animations, sound, and video) and create hyperlinks.

balance A design principle that refers to the distribution of optical weight in the screen layout.

bandwidth The size (measured in the amount of data that can be transmitted in a set period of time) of the connection between two computers on a network.

beta testing The final functional test of a software program before release. It involves selected potential users that could number in the thousands. A goal of the beta test is to get feedback from as wide a variety of users on as many different computer configurations as possible. See also *alpha testing*.

binary The system used to represent the coding of instructions and data for a computer.

bitmap An image that is displayed as an array of pixels.

bit A binary digit—the smallest unit used to represent the coding of instructions and data for a computer.

book metaphor The metaphor used by some authoring programs wherein the multimedia title is thought of as a book, and each screen display is a page in the book.

browser A program, such as Netscape Navigator or Microsoft Internet Explorer, that allows a user to display HTML-developed Web pages.

budget A financial plan used in project management that affects decisions, such as which elements are used in a multimedia title, and acts as a control mechanism to evaluate the status of a project.

bundling Distribution of a multimedia title with some other product such as a new computer or an upgrade kit.

card stack metaphor The metaphor used by some authoring programs wherein the multimedia title is thought of as a stack of cards, with each card being a screen display.

CD See *compact disc*.

CD Audio See *compact disc–digital audio*.

CD-DA See *compact disc–digital audio*.

CD-E See *compact disc–erasable*.

CD-I See *compact disc–interactive*.

CD-R See *compact disc–recordable*.

CD-ROM See *compact disc–read-only memory*.

CD-ROM drive A device used to play CD-ROMs.

CD-ROM extended architecture (CD-ROM XA) An extension of the CD-ROM format that allows for interleaving data to enhance the playback of sound and video. It provides the same benefit of better synchronized sound and pictures as

CD-I, but it works with the computer rather than dedicated equipment attached to a TV.

CD-ROM XA See *CD-ROM extended architecture*.

CD-WO See *compact disc–recordable*.

cel animation A type of animation that is based on the changes that occur from one frame to another.

censorship An issue related to who will regulate the contents of a CD title, especially concerning pornography, racism, and violence.

CGI See *Common Gateway Interface*.

client A network desktop computer that allows users access to the programs and data on a server.

codec Programs used for video compression and decompression.

color depth The range of colors available for each pixel (8-bit, 256; 16-bit, 65,000; 24-bit, 16.7 million).

Common Gateway Interface (CGI) A coding standard that allows programmers to write applications for the Web. A typical application is one that allows the user to search a database.

compact disc (CD) A popular medium for storing and delivering multimedia titles because of the large capacity (650 MB) of a CD.

compact disc–digital audio (CD Audio) The format for storing recorded music in digital form, as on CDs that are commonly found in music stores. Using compression techniques, CD Audio discs can hold up to 75 minutes of sound. Also referred to as *CD-DA*.

compact disc–erasable (CD-E) A CD format that allows a user to update information on the disc and free up disc space by erasing unneeded data. Erasable CDs are especially beneficial to multimedia developers and those needing to exchange data, archive large amounts of data, and back up data stored on hard drives.

compact disc–interactive (CD-I) A platform-specific format requiring a CD-I player, with a proprietary operating system, attached to a television set. One of the benefits of CD-I is its ability to synchronize sound and pictures on a single track of the disc.

compact disc–read-only memory (CD-ROM) The format for storing data, including video and audio, in digital form on a compact disc. CD-ROMs can hold up to 680 MB of data.

compact disc–recordable (CD-R) A file format that allows single CDs to be produced using a desktop compact disc recorder. Also referred to as *compact disc–write once* or *CD-WO*.

compact disc recorder Laser-based system that creates one CD at a time—referred to as a *one-off*. Depending on the model, compact disc recorders can create discs for all the major CD formats, including multisession discs that can be written to more than one time.

compact disc–write once See *compact disc–recordable*.

compression The process of reducing the size of a file, such as a graphic image, by eliminating redundant data.

cooperative advertising A marketing technique whereby the retailer and the publisher share the cost of advertising. This involves promoting the multimedia title and the retailer in the same advertisement in order to generate demand for the title and traffic for the store.

copyright Laws designed to protect intellectual property rights and provide potential monetary rewards for inventiveness and hard work. The ease with which material can be copied, digitized, manipulated, incorporated into a title, and delivered to a mass market has prompted a concern about the adequacy of copyright laws as they apply to the multimedia industry.

demand The potential number of sales for a multimedia title.

demographics A way to describe potential audiences, including their location, age, sex, marital status, education, and income.

derivative works Creative works based on an original work, such as translations, abridgments, adaptations, or dramatizations.

design principles Features of a multimedia title that relate to the appearance (What should the screen look like?) and the interaction (How does the process work?) of the title.

developer A person, company, or organization that creates multimedia titles.

development system A high-end computer system used to create multimedia titles.

digital camera A camera used to capture still images that can be transferred directly to a computer.

digital video disc (DVD) A CD with a storage capacity of up to 17 GB. This allows full-length movies with different audio tracks (to accommodate various languages), and even different versions of the same movie (PG, PG-13, R) to be available on one disc. The technology involves increasing the data density by reducing the size of the pits and lands, and providing double-layered and double-sided discs.

drawing program Tools used to create draw-type graphics that provide for freehand as well as geometric shapes and are useful in creating designs in which precise dimensions and relationships are important.

draw-type graphic An image that is represented as a geometric shape made up of straight lines, ovals, and arcs.

DVD See *digital video disc*.

edutainment A multimedia title that is both educational and entertaining.

electronic slide show A computer-based multimedia presentation similar to a presentation using overhead transparencies or slides.

elements Components of a multimedia title, including text, sound, graphics, animation, and video.

exploratory navigation A navigation scheme providing little structure or guidance, relying instead on user interaction—usually by clicking on objects displayed on the screen. Many games, directed at both children and adults, use some form of exploration.

external storage Devices that are connected to a computer and provide additional storage space needed to accommodate the large files that are created when developing multimedia titles.

feature creep A phenomenon in the development process in which the specifications of a multimedia title are changed by adding new features. This is common as companies try to enhance a product as it is being produced. It can result in added costs and missed deadlines.

feedback A process whereby a user obtains some indication that an action has been recognized by the computer, such as a sound effect when the user clicks on a button.

font An entire group of letters and characters of a specific design.

fps See *frames per second*.

frames per second (fps) A measurement of the number of still images (frames) that are displayed onscreen in one second.

frequency The distance between peaks in a sound wave pattern; the greater the distance, the lower the pitch.

GB See *gigabyte*.

GIF See *Graphics Interchange Format*.

gigabyte (GB) One billion bytes.

goals Broad statements of what a multimedia project will accomplish, as contrasted with objectives, which are more-precise statements. An example of a goal statement is: *Be the leader in educational CDs*.

graphic A multimedia element such as a drawing, photo, or piece of clip art. In CD labeling, it refers to the artwork placed on the surface of the CD for promotional or informational purposes.

Graphics Interchange Format (GIF) A standard graphics file format for the Web which automatically compresses graphic images when they are created.

graphics program Tools used to create and edit graphic images.

Green Book The CD-I (compact disc–interactive) format specifications. *Green* refers to the color of the binding of the document in which the specifications were first published.

helper application A program used to view elements not viewable with a browser alone. Helper applications display an element (such as a video clip) in a separate window on the user's monitor.

hertz (Hz) A measurement of a sound pattern frequency, with a recurring pattern every second equaling 1 hertz.

home page A primary page for a Web site. The home page is generally the anchor that is used as a reference point by the site developer. All other pages of the site link to this page and usually contain a navigation button that takes the user directly to the home page.

HTML See *Hypertext Markup Language*.

hyperlinking The process of linking elements so that the user can jump from one part of a multimedia title to another.

Hypertext Markup Language (HTML) A programming language specifically designed for the Internet. It allows multimedia to be incorporated into the Internet by providing hyperlinking and the ability to use sound, animation, video, and graphics on a Web page.

hypertext The linking of elements that allows a user to jump to another part of a multimedia title.

Hz See *hertz*.

icon-based program An authoring tool that uses a flowchart scheme and icons to represent particular events, such as playing a sound.

image file The result of the premastering process that creates an exact image of what will be placed on a finished CD, including the data, filenames and directories, error detection and correction routines, indexes, and programs (such as an install program).

image-editing program Tools specifically designed to manipulate graphic images, such as Photoshop.

individualization The ability of a multimedia title to address different learning styles and needs.

informational kiosk A computer-based information distribution system such as those found in shopping malls and museums.

Integrated Services Digital Network (ISDN) A technology that enables digital signals to be transmitted over phone lines, thereby increasing the speed at which multimedia elements are delivered from one computer to another.

interactive A feature of a multimedia title that allows the user to control some aspects of the title.

interactivity The ability of a multimedia title to allow for user control.

Internet A network of networks that forms a vast communications system linking computers around the world. Developed by the government and various academic institutions, the Internet has evolved into a major commercial, research, entertainment, education, and communications network.

intranet An internal network set up by companies and organizations to facilitate communications among employees, customers, vendors, and selected others.

ISDN See *Integrated Services Digital Network*.

ISO 9660 International Standards Organization specifications that allow CDs to be played across various computer platforms including Macintosh, Windows-based, and UNIX-based computers. They include standards for the maximum number of levels in a directory structure (eight), and filename specifications (the DOS eight-dot-three convention).

Java A powerful programming language developed specifically for the WWW by Sun Microsystems. Similar to C++ (a popular language used to develop various applications such as word processing programs and games), Java is used to extend the functionality of HTML.

jewel box A clear plastic case slightly larger than a CD that has become standard CD packaging, because of its relatively low cost, superior protection, and marketing features.

Joint Photographic Experts Group (JPEG) A standard graphics file format for the Web which automatically compresses graphic images when they are created by eliminating redundant information in the image.

JPEG See *Joint Photographic Experts Group*.

KB See *kilobyte*.

kilobyte (KB) One thousand bytes.

kiosk A computer-based system that allows transactions, such as airline self-ticketing systems, or provides information, such as those systems found in shopping malls.

labeling Applying to a CD, through a silk-screening process, the name of the title, the name and logo of the company that developed the title, copyright information, and promotional graphics.

leading Involves influencing others to achieve the project goals. This is extremely critical in multimedia development, because each member of the team brings highly technical and specialized skills to the project.

"look and feel" The tone, approach, metaphor, and emphasis of a multimedia title.

lossless compression A type of file compression that preserves the exact image throughout the compression and decompression process. Results in a better-quality image and a larger file size than lossy compression.

lossy compression A type of file compression that eliminates some of the data in the image during compression and decompression. Results in a poorer-quality image and a smaller file size than lossless compression.

mastering The production process of taking a premastered image file and creating a master disc that is used as a mold to create a *stamper*, which in turn is used to create the finished CDs.

MB See *megabyte*.

megabyte (MB) One million bytes.

metaphor A comparison of how a multimedia title or authoring program works and the user's frame of reference. For example, the authoring program ToolBook uses a book with pages as the metaphor for a multimedia title with screens. That is, each ToolBook page is a screen in the multimedia title.

MIDI See *Musical Instrument Digital Interface*.

milestones Significant accomplishments in the multimedia development process, such as the completion of a prototype for usability testing. Whether or not a milestone is achieved on schedule is critical to management's control of the project.

millisecond (ms) One-thousandth of a second.

MMX See *Multimedia Extensions*.

modeling The process of creating objects and scenes as the first step in creating a 3-D animation.

monitor The display unit of a computer.

morphing The process of evolving one image into another via a series of images to create a special animation effect.

Motion Picture Experts Group (MPEG) The organization that developed a standard for the compression of a series of images, which works by recording the changes in an image from key frame to key frame.

movement A design principle having to do with how the user works through the elements on the screen, including where the eyes are initially drawn and how the eyes move around the screen.

MPC See *Multimedia Personal Computer (MPC) specifications*.

MPEG See *Motion Picture Experts Group*.

ms See *millisecond*.

multimedia A computer-based interactive communications process that incorporates text, graphics, sound, animation, and video.

Multimedia Extensions (MMX) A technology developed by Intel to boost the performance of video, audio, communications, and graphics on Intel processors, which traditionally have not been very adept at handling these multimedia tasks.

Multimedia Personal Computer (MPC) specifications A computer configuration that is a standard for Windows-based computers designed to play multimedia titles.

multimedia presentation A computer-based presentation that is delivered by a presenter to an audience. It may include text, sound, graphics, animation, and video along with hyperlinking capabilities.

multimedia title A multimedia application such as a reference, entertainment, or educational product.

multisession disc A Photo CD or CD-R that can be written to more than one time.

Musical Instrument Digital Interface (MIDI) A standard format that enables computers and electronic musical instruments to communicate sound information.

nature The essence of a multimedia title, suggesting a theme that the interactive design would need to complement and reinforce. For example, if the title is an astronomy application in which students study the night sky, the view could be through a telescope that the user manipulates with the mouse.

network An interlinking of two or more computers. Networks are often configured with a more powerful computer, called a *server*, that controls the network and provides a large storage capacity. The other computers on the network, called *clients*, allow users access to the programs and data on the server.

nonlinear The feature of a multimedia title that allows the user to determine what content is delivered, as well as when and how it is delivered.

objectives Precise statements of what the project will accomplish, as contrasted with goals, which are more-general statements.

OCR See *optical character recognition*.

OEM See *original equipment manufacturer*.

one-off The process of using a compact disc recorder to create one CD at a time.

online distribution Delivery of multimedia titles via the Internet.

online A process for delivering multimedia, utilizing telecommunications and the Internet.

optical center A point somewhat above the physical center of the screen in a balanced design.

optical character recognition (OCR) A program that is used with a scanner to capture text that can be edited using a word processor.

optical weight The ability of an element (such as a graphic, text, headline, or subhead) to attract the user's eye. Each element has optical weight as determined by its nature and size. The nature of an element refers to its shape, color, brightness, and type. For example, a stunning color photograph of Mount Everest would have more weight than a text block of equal size.

organizing In project management, refers to making sure the necessary resources, especially personnel, are available. A critical aspect of multimedia development is the formation of the development team. This involves assessing the current talent in light of the requirements for a particular project.

original equipment manufacturer (OEM) In distribution of a multimedia title, the developer may bundle the CD with some other product such as a new computer or an upgrade kit. The advantage for the publisher is that essentially no marketing costs are involved other than negotiating with the hardware manufacturer. This is a quick way to get a product out to a new market in hopes of establishing brand identification and follow-up sales of new versions.

packaging The final step in the production of CDs in which decisions are made on the type of package, such as jewel boxes, and the promotional materials and documentation to include with the package.

paint program Tools used to create bitmap images.

path animation A type of animation created by moving an object along a predetermined path on the screen.

peripherals Hardware devices that link externally to a computer and are useful in developing multimedia titles. Examples are a scanner, external storage equipment, and a CD recorder.

Photo CD Developed in 1992 by Kodak as a means of storing and viewing photos, slides, and film transparencies. The process involves scanning the photos with a high-end scanner, compressing the images, and writing them to a CD. More than a hundred photos can be stored on a CD.

pits and lands The recessed and level areas on a compact disc, created when the finished CD is replicated. The pits and lands represent the digital coded data. When a laser beam is passed over the disc, the light is reflected by the lands and not by the pits. The reflected light is read by a sensor, and a signal is sent to the computer, which translates the signal into binary code.

pixel A picture element—the smallest unit a monitor can display.

planning In project management, refers to how to achieve the objectives of an individual title. Three important planning aspects are tasks, budgets, and schedules.

playback system A computer that is configured to play multimedia titles, as contrasted with a development system used to create multimedia titles.

plug-in A program, such as Shockwave, that is used to view elements not viewable with a browser alone. A plug-in displays an element as though it were part of the browser.

point size A unit of measure for measuring fonts. There are 72 points in 1 inch.

point-of-purchase Promotional materials, such as displays, that are placed in a retail store.

premastering The process of creating an exact image of what will be placed on a finished CD, including the data, filenames and directories, error detection and correction routines, indexes, and programs (such as an install program).

press kit A technique to generate publicity whereby multimedia publishers prepare and distribute to selected reviewers press kits containing product information and a full version (not a demonstration version) of the title.

privacy Laws that include two issues important to multimedia developers. First, revealing embarrassing facts about an individual that would be considered offensive to a reasonable person and where there is no sufficient cause (such as a news story) for the disclosure may violate privacy laws. Second, placing a person in a false light which causes undue stress on the individual could also constitute a violation of privacy.

processor The brain of a computer system—it controls the computer's operations and performs calculations.

programming language Software tools used to create a variety of applications, including multimedia titles. Programming languages are more flexible than authoring tools but may not be as easy to use.

public domain Materials that have no copyright are considered in the public domain and can be used without permission. Either no copyright was issued (such as is the case with certain government-generated materials), the copyright has expired (for some works this is 75 years after publication), or it was not renewed. There may be legal considerations when using public domain materials, especially those related to derivative works, trademarks, and people.

publicity Stories that appear in various news media such as magazines, newspapers, and trade publications. The stories often appear as reviews of new multimedia products and can be favorable or unfavorable.

publisher Those involved in the marketing and particularly the distribution of multimedia titles. In many cases, the same company both develops and publishes a title.

RAM See *random access memory*.

random access memory (RAM) The area of a computer where data is temporarily stored while processing takes place.

read-only memory (ROM) The area of the computer where information is permanently stored, such as the instructions needed to begin the process of starting a computer.

Red Book The CD Audio (compact disc–digital audio) format specifications. *Red* refers to the color of the binding of the document in which the specifications were first published.

rendering The final step in creating a 3-D animation that involves giving objects attributes such as colors, surface textures, and degrees of transparency.

replication The process in the manufacture of a CD wherein a master disc is used to produce a quantity of CDs.

resolution A measurement of the number of pixels displayed on a monitor. The higher the number of pixels, the greater the resolution and the better the image quality.

retailer A store that sells primarily to the general public.

Right of Publicity The rights of individuals; this is a legal basis for requiring permission and/or payment for using a person's name, image, or persona.

ROM See *read-only memory*.

sampling The process of digitizing sound by recording parts of the sound every fraction of a second.

sans serif A font in which there are no lines or curves (serifs) that extend from the strokes of letters.

scanner A hardware device used to capture images that are saved in a graphics file format for use in a multimedia title.

schedule A list of a project's beginning and ending dates and milestones. It is useful in coordinating the efforts of team members by showing how the various activities relate to one another.

script Programming code that is written by a multimedia developer or generated by an authoring program.

sequential navigation A navigation scheme that takes the user through a more or less controlled, linear process. An example is a game with a story line that has a beginning, middle, and end.

serif A font style in which a line or curve (a serif) extends from the strokes of letters.

server A network computer that is used to control the network operations and provide a large storage area. A server is generally more powerful than the client computers on the network.

service provider A company that specializes in developing multimedia titles. A multimedia service provider can essentially become the multimedia arm of a company and take over the primary function of managing a project.

shelf space The area on a retail store shelf that is available to a particular CD title.

sound card A hardware device installed in a computer that is used to digitize sound from some external source such as a videotape player, CD, or microphone.

specifications A list of what will be included in each screen of a multimedia title, including the arrangement of each element and the functionality of each object (for example, what happens when you click on the button labeled *Next*).

SRP See *suggested retail price*.

stand-alone titles Those multimedia applications that are meant for use by individuals in a one-on-one situation.

storyboard A representation (often in the form of hand-drawn sketches) of what each screen of a multimedia title will look like and how the screens are linked. The purpose of a storyboard is to provide an overview of the project, act as a guide (road map) for the programmer, illustrate the links among screens, and illustrate the functionality of the objects.

street price The price that a consumer generally pays for a particular CD title. It is often less than the manufacturer's suggested retail price.

stand-alone, 95–96
storyboard and navigation and, 127–130
target audience and, 120
testing, 133–134
Multimedia upgrade kit, 36
Multiple Zones International, 209
Multisession disc, 187
Musical Instrument Digital Interface (MIDI), 74
Myst, 13, 209

Nature, 142
Navigation, 127–130, 149
 exploratory, 130
 sequential, 129–130
 topical, 130
Navigation bar, creating, 290–296
NEC Technologies, 30
Netscape, 222, 223, 232
Network, 220
New Media Express, 208
Nintendo Game Boy, 12
Nondisclosure agreement, 134, 135
Nonlinear, 5

Objectives, 119–120
Object motion options, 279
Objects, 268
OCR program. *See* Optical character recognition program
OEM marketing, 209
Online, 18–19
Online distribution, 212–213
Optical center, 146
Optical character recognition (OCR) program, 53
Optical weight, 142
Organizing, 171–174

Packaging, 192–194
PageMill, 222
Paint, 57, 59
Paint program, 57
Palette, 264–268
Panel, 264–268
Parent's Choice Foundation, 207
Partnership, 211
Path animation, 76, 320
Path Editor, 322–324
Pause control, 336

PC Connect, 209
PCX, 59
PC Zone, 209
J.C. Penney, 14
Personal computer. *See* Multimedia personal computer
Persuasion, 96
Phantasmagoria, 252
Philips Consumer Electronics, 30, 185, 251
Photo CD, 186–187
PhotoDisc, 62
PhotoDisc Web site, 228–231
Photoshop, 59
PICT, 59
Pits and lands, 183–184
Pixel, 32–33, 55
Planning, 169–171
Playback head, 336
Playback system, 28, 31–36
 target, 124–125
Play Time, 278
Plug-in, 232
Point-in-time view, 281
Point-of-purchase, 202, 203
Point size, 49
Pop-up message, 52
PostScript font, 51, 60
Power Mac, 38
PowerPoint, 96, 97
Premastering, 188–189
Premiere, 88
Prentice-Hall, 214
Presentations. *See* Multimedia presentation
Press kit, 205
Pricing strategy, 205–206
Privacy, 248
Processor, 31, 38
 new generation, 250
Production, 182
Product strategy, 201–202
Program manager, multimedia project team and, 173
Programmer, multimedia project team and, 173
Programming languages, 104–109
Project manager, multimedia project team and, 173
Project schedule, 171, 172
Promotion strategy, 202–205
Public domain, 245–246
Publicity, 202–205
Publishers, 200

Quest, 109
QuickTime, 81, 87, 331
QuickTime for Windows, 87

Random-access memory (RAM), 31–32
Random House New Media, 211
Rapid Assault, 209
Reader's Digest, 211
Read-only memory (ROM), 32
Rebel Assault, 209
Recreation, multimedia and, 15–16
Red Book, 185
Reference, multimedia and, 15
Regency Script, 46, 47
Rendering, 77, 78
Rentals, 210–211
Replication, 189–192
Resolution, screen, 32, 33
Retailer, 208
Retail price, suggested, 205
Right of Publicity, 247
ROM, 32

Sample rate, 71
Sample size, 71
Sampling, 35, 71–72
San serif font, 46–49
Scanned image, 63–64
Scanner, 38–39, 63
Scene layer, 269
Scene menu, 263, 273, 282, 288, 321, 329
Schedule, 171, 171
Scholastic, 211
Screen capture program, 64
Screen design
 balance and, 140–144
 movement and, 146, 147
 unity and, 144–145
Screen resolution, 32, 33
Script, 104
Scripting, 104
Scroll bar, 52, 264
Search, 306
Segment, 320
Sequential navigation, 129–130
Serif font, 46–49
Server, 220
Service provider, 164–165. *See also* Multimedia service provider

SESAC, 246
7th Guest, 13
Shelf space, 201
Shockwave, 109, 222, 231, 232
Sierra On-Line, 165, 252
Sim City 2000, 209
Software Publishers Association, 30
Software Publishing, 96
Sony, 185, 186
Sound, 70–73, 286
 adding to presentations, 317–319
Sound card, 72
Sound-editing program, 73
Span-time view, 281
Specialty retailer, 208–209
Specifications, 124–127
Spiegel, 14
Sports Illustrated for Kids, 209
Sports Illustrated Multimedia Almanac, 211
Stand-alone title, 95–96
StarPress Multimedia, 211
Start Time, 278
Status bar, 264
Still image, sources, 63
Stock photograph, sources, 62–63
Storyboard, 127–130, 260
Suggested retail price, 205
Sun Microsystems, 228
Superstore, 208–209
Super VGA (S-VGA), 32
Symmetrical balance, 142–143

T1 line, 225
T3 line, 225
Tabs, 151
Tagged Image File Format (TIFF), 59
Tags, 225
Target, 14, 208
Target audience, 120, 200
Target market, trends and, 251
Target playback system, 124–125
Task analysis, 169
Telecommunication, 18–19
Template, 156, 269
 creating, 270–275
Template menu, 270
Template Options window, 271

Testing, 133–134
    alpha, 134
    beta, 134
Text, 46–51
    embedded, hyperlinking using, 312–314
    software for creating and editing, 53
Text alignment options, 278
Text-intensive title, 51–52
Text menu, 264, 277
Thumbnail, 96, 97
TIFF, 59
Time-based program, 102–103
Timeline, 334–339
Time scale button, 336
Times Mirror, 211
Times New Roman, 51
Time tab, 274
Title, text-intensive, 51–52
Title bar, 262, 306
To End of Scene, 278
Tone, multimedia titles and, 121
Tool bar, 264
ToolBook, 98, 99, 100, 163
    programming language, 105–106
Tool palette, 264–268
Topical navigation, 130
Toys R Us, 208
Trademark, 246–247
Training, multimedia and, 14, 15
Transactional kiosk, 213
Transactions, titles allowing, 150
Transition tab, 274
TrueType font, 51, 60
TRW, 15
Tweening, 76
2010 Media, 246, 247
Typeface. *See* Font
Type style, 50

Underline, 50
Undo command, 263
Unity, screen design and, 144–145
UNIX, 223
Upgrade kit, 36
Usability tester, multimedia project team and, 173
User interface, 127

Value-added, 206
Vector graphics, 54
Video, 82–89
    compression/decompression, 85–87
    digitizing, 82–83
    file size and, 83–85
    software for capturing and editing, 88–89
Video archives, organizations providing stock footage, 247
Video card, 32–35, 38
Video clip
    editing, 88
    playing, 331–334
Video for Windows, 87, 331
Video Graphics Array (VGA), 32
View menu, 263, 270, 272, 288
View selector, 264
Virtual reality (VR), 81
Virtual Reality Modeling Language (VRML), 233
Visual Basic, 104
Voyager Company, 249
VR. *See* Virtual reality
VRML. *See* Virtual Reality Modeling Language

Wal-Mart, 208
Warping, 80
Warren Miller Films, 245
Web-based authoring tools, 109
Web browser, 231–232
Web page, 221, 226, 230
    layout, 234–237
Web site, 237–238
Web specialist, 223
Wholesaler, 206–207
Window menu, 264, 267
Windows, 5, 29, 37, 51, 54, 59, 60, 64, 87, 96, 124–125, 148, 163, 223
    MPC specifications, 30
Windows 3.1, 262
Windows 95, 57, 125, 134, 262
The Winogradsky Co., 246
Work area, 264
World Wide Web (WWW), 14, 19, 59, 124, 220–223
    animation and, 232–233
    authoring tools, 109
    multimedia and, 223–233
WWW. *See* World Wide Web

XObjects, 108
Xtras, 109

Yellow Book, 186

Zany Brainy, 208
Zoom tools, 336

CREDITS

FIG.	SOURCE
1.6	Photo by Phil Matt.
1.7	Photo © Bob Daemmrich.
2.2	The MPC logo is a registered trademark.
2.8	Photo by Phil Matt.
3.15	Courtesy of Corel, Inc.
3.16	Courtesy of Hewlett-Packard.
4.4	Photo by Phil Matt.
5.7	Courtesy of Jennifer Fulton.
5.11	Courtesy of Jennifer Fulton.
5.12	Courtesy of Jennifer Fulton.
6.1	Courtesy of Brøderbund Software, Inc.
6.2	Courtesy of Sherluck Multimedia, Ltd.
6.3	Courtesy of Humongous Entertainment.
7.4	Courtesy of Reilly Jensen.
7.5	Courtesy of Sue Willoughby.
7.7	Courtesy of Sue Willoughby.
7.15	Courtesy of Corbis, Inc.
7.16	Courtesy of Corbis, Inc.
9.5	First photo by Philippe Plailly/Science Source/Photo Researchers, Inc.; all others by Robert Holmgren.
9.6	Photo by Scott Goodwin.
9.7	Photo by Phil Matt.
9.8	Photo by Scott Goodwin. Insert cover © CD Access, Inc.
10.1	Courtesy of Brøderbund Software, Inc.
10.2	Courtesy of Egghead, Inc.
10.3	Photo by Phil Matt.
10.4	Reprinted from *PC/Computing,* November 1996. Copyright © 1996 Ziff-Davis Publishing Company.
10.6	Photo by Scott Goodwin.
11.3	Courtesy of Joe Burns.
12.3	Courtesy of Pioneer Electronics.

Hypertext, 50, 96
Hypertext Markup Language (HTML), 221, 223, 225–228
Hz. *See* Hertz

IBM, 30, 209
Icon-based program, 101–102
Idea Factory, 245
Illustrator, 53, 57
Image database, 132
Image-editing program, 57–59
Image file, 188
Image size
  bitmap graphics and, 55–56
  digitized video and, 84
Index, 306
Individualization, 11
Informational kiosk, 213
Ingram Micro, Inc., 206, 212
Instructional designer, multimedia project team and, 173
Integrated Services Digital Network (ISDN), 224
Interactive design, 147–153
  guidelines, 154–157
Interactive Home Systems (IHS), 244
Interactive multimedia, developing, 116–117
  creating phase, 131–133
  planning phase, 117–130
  testing phase, 133–134
Interactivity, 5–7, 140
Interface designer, multimedia project team and, 173
InterMedia Active Software, Inc., 211
International Standards Organization (ISO), 186
International Thompson, 214
Internet, 19, 220–223
  limitations, 224–225
  multimedia titles and, 249–250
Internet Explorer, 222, 223
Internet service provider, 222
Intranet, 225
Irwin, 214
ISDN, 224
ISO 9660, 186, 188, 189
Italic, 50

Jack Nicklaus Golf, 209
Java, 228, 229
Jewel box, 192–193
Joint Photographic Experts Group (JPEG), 59, 85, 224
  compression standards, 85–87

Kilobyte (KB), 36
Kiosk, 17–18
  informational, 213
  transactional, 213
Kiosk-based multimedia, 213–214
K-Mart, 208
Kodak, 186–187

Labeling, 192
Layer, 268, 269
Leading, 174–175
Living Books, 165, 201, 211
"Look and feel," 121–123
Lossless compression, 85
Lossy compression, 85

Macromedia, 231, 258, 331
Mac Zone, 209
Magic Carpet, 209
Magic School Bus Explores the Solar System, The, 211
Main section, 306
Management, multimedia projects and, 165–175
Market, target, trends and, 251
Marketing, multimedia and, 9–10, 14, 200–201, 211–212
Mastering, 189–192
Math Blaster, 15
Media Mosaic, 211
Media professional, multimedia project team and, 173
Megabyte, 28
Memory, 31–32, 38
  random-access, 31–32
  read-only, 32
Menu bar, 262–263, 306
Merchant, The, 14
Mercury Magazine, 211
Meredith, 211
Metalization, 190
Metaphor, 96, 154
  book, 98
  card stack, 98
  multimedia titles and, 122

Microphone, 41
Microsoft, 30, 165, 186, 202, 211, 212, 244, 250
Microsoft Network, 212
Microsoft Windows. *See* Windows
Micro-warehouse, 209
MIDI, 74
Milestones, 171
Millisecond (ms), 36
MMX, 250
Modeling, 77, 78
Modem, 224
Monitor, 32–35, 38
Morphing, 79–80
Mosaic, 221
Motion Picture Experts Group (MPEG), compression standards, 85–87
Movement, screen design and, 146, 147
MPC logos, 30
MPC specifications, 30
  levels, 31
MPEG compression, 85–87
MS-DOS, 189
Multicom, 211
Multimedia, 5
  censorship issues and, 248–249
  copyright issues and, 244–248
  delivering, 16–19
  development, management issues and, 162–165
  elements, 8
  examples, 11
  growth, 8–11
  inappropriate use, 19–20
  interactive. *See* Interactive multimedia
  privacy issues and, 248
  World Wide Web and, 223–233
Multimedia company, 165
Multimedia computer system
  development, 28, 37–40
  playback, 28, 31–36
Multimedia Extensions, 250
Multimedia industry, trends, 249–252
Multi-Media Music Web site, 231
Multimedia PC Marketing Council, 30

Multimedia personal computer, 29–30
  typical configuration, 29
Multimedia Personal Computer (MPC) specifications, 30
  levels, 31
Multimedia presentation, 14–15, 94–95, 258
  adding sound, 317–319
Multimedia project
  forming the team and, 171–174
  management process and, 165–175
  organizing resources and, 171–174
  planning, 169–171
Multimedia service provider, 164–165
  contract, 167
  promotional packet, 166
  rate card, 168
Multimedia title, 5
  authoring, 132–133
  budget, 170
  concept development and, 117–118
  consumer, marketing, 200–201
  content development and, 131–132
  decrease in price, 9, 10
  design principles and, 140–146
  distributing on CD-ROM, 200–212
  distributing online, 212–213
  distribution alternatives, 209–211
  distribution strategy, 206–209
  goals and objectives and, 119–120
  growth, 8–9
  interactive design and, 148–150
  Internet and, 249–250
  kiosk-based, 213–214
  "look and feel" and, 121–123
  major categories, 12–16
  pricing strategy, 205–206
  product strategy for, 201–202
  promotion strategy, 202–205
  specification development and, 124–127

Index

premastering, 188–189
production process, 188–195
  costs, 194, 195
CD-ROM drive, 35–36
CD-ROM title, 17. *See also* Multimedia title
CD-ROM XA, 186
Cel animation, 75–76
Censorship, 248–249
CGI. *See* Common Gateway Interface
Chart menu, 264
Claris, 165
Client, 220
Clip art, sources, 62–63
CMEA, 207
CNN ImageSource, 247
Codecs, 85
Collage Plus, 64
Color depth, 55–56
  digitized video and, 84–85
Color palette, 271
ComicsCarToon, 46, 47
Common Gateway Interface (CGI), 228
Communication, corporate, multimedia and, 14–15
Compact disc (CD), 17, 182–187
  formats, 185–187
  pits and lands, 183–184
  reading, 184
  spiral track, 185
  storage capacity, 28, 182
  structure of, 183–184
Compact disc–erasable (CD-E), 251
Compact disc–read-only memory (CD-ROM), 35, 186
Compact disc–recordable (CD-R), 187
Compel, 96, 133
Complete Baseball, 212
Compression, 224
Compton's Encyclopedia, 209
Compton's NewMedia, 30
CompUSA, 208
Computer. *See* Multimedia personal computer
Computer City, 208
Consistency, interactive design and, 155–156
Consumer titles, marketing, 200–201

Content
  interactive design and, 150–153
  trends, 251–252
Content expert, multimedia project team and, 173
Contents section, 306
Control panel, 265
Cooperative advertising, 202, 203
Copyright, 244–248
Copyrighted materials, acquiring rights, 245
Copyrighted music, organizations granting rights, 246
Copyright Music and Visuals, 246
Corbis Corporation, 63
CorelDRAW, 53
Corel Gallery 2, 63
Corel Stock Photo Library, 63
Corporate communication, multimedia and, 14–15
Creative Video, 247
Critter, 46, 47

Dandy Dinosaur, 211
Davidson, 207
Decorative font, 46
Defense Department, 220
Demand, 208
Demographics, 120
Derivative works, 246–247
Design, interactive, 147–153
  guidelines, 154–157
Design principles, 140–146
Design template, 156
Developer, 200
Development, trends, 251–252
Development system, 28
Diamond Multimedia Kit, 209
Digital camera, 40, 41
Digital video disc (DVD), 250, 251
Digital Video Editor, 334
Direct mail, 209
Director, 102, 222, 231, 233, 234, 258
  programming language, 107–108
  third-party modules, 108–109
Director Player, 331
DirectX, 250

Distribution, 182
Distribution strategy, 206–209
Drawing program, 57
Draw-type graphics, 54–55
Drop-down box, 52
DVD. *See* Digital video disc

Edit menu, 263, 289, 293, 296, 298, 302, 327, 328
Education, multimedia and, 13
Edutainment, 15
Egghead, 208
Ehrich Multimedia, 211
Electronic slide show, 96, 97
Embedded text, hyperlinking using, 312–314
Emphasis, multimedia titles and, 122–123
Encarta, 15
Encyclopedia, multimedia, 6–7
End-user, growth in multimedia and, 10–11
Enter, 279
Entertainment, multimedia and, 12–13
Exit, 279
Explorapedia, 122
Exploratory navigation, 130
External storage device, 39–40

Fabulous Footage Inc., 247
Feature creep, 171
Federal Aviation Administration (FAA), 81
Feedback, 134, 157
File menu, 197, 282, 288, 338
File size, bitmap graphics and, 55–56
Film archives, organizations providing stock footage, 247
Filter, 60
Fine art, sources, 62–63
Flowchart, 6, 261
Follette Campus Resources, 214
Font, 46–51
  sans serif, 46–49
  serif, 46–49
Font dialog box, 277
Food & Wine Tasting, 211
Four Palms Web site, 231
fps. *See* Frames per second

Frame rate, 83
Frames per second (fps), 74
Frequency, 70
Front Page, 222
Functionality, multimedia titles and, 125
Funk and Wagnalls, 15

Games. *See* Entertainment
Gates, Bill, 244
GIF, 59, 224
Goals, 119
Go Back, 306
Golf Digest, 209
Graphic artist, multimedia project team and, 173
Graphics, 54–56
  adding motion, 316–317
  adding with hyperlinks, 314–316
  bitmap, 55–56
  draw-type, 54–55
  vector, 54
Graphics image, sources, 62–64
Graphics Interchange Format (GIF), 59, 224
Graphics program, 53, 57–59
  features, 59–62
Graphics tablet, 60, 61
Green Book, 186

Hardware, trends, 250–251
The Harry Fox Agency Inc., 246
Harvard Graphics, 96
Helper applications, 232
Help feature, 305–307
Help menu, 264, 306
Helvetica, 51
Hertz (Hz), 70
Highlight, 286
Hijack Pro, 64
History, 306
Hold, 279
Holiday Inn, 15
Home page, 237
Hotword, 50
HTML. *See* Hypertext Markup Language
HyperCard, 98
HyperCard stack, 99
Hyperlinking, 52, 96, 105, 221, 226, 237–238
  adding graphics, 314–316
  using embedded text, 312–314

INDEX

ABCNEWS VideoSource, 247
Acquisitions specialist, multimedia project team and, 173
Action, 286
Action! 3.0, 258–261
 adding body text, 302–305
 adding graphics with hyperlinks, 314–316
 adding motion to graphics, 316–317
 adding scene, 282–284
 adding sound to presentation, 317–319
 adding text and linking it to another scene, 298–302
 adding text to scene, 275–280
 copying group of objects, 296–298
 copying scenes, 288–290
 creating animations, 319–331
 creating buttons and linking scenes, 284–288
 creating navigation bar, 290–296
 creating scene, 268–269
 creating template, 270–275
 developing title with, 262–264
 editing movie, 334
 getting help while using, 305–307
 hyperlinking using embedded text, 312–314
 palettes and panels, 264–268

playing presentation of scene, 281
playing video and animation clips, 331–334
printing presentation, 307–308
saving presentation of scene, 282
working with timeline, 334–339
ActiveX, 109
Adobe, 53, 57, 88, 96
Advertising, cooperative, 202, 203
Alpha testing, 134
Alpine, 211
Analog wave pattern, 70, 71
Animation, 74–80
 cel, 75–76
 creating, 319–331
 nonlinear, 320
 path, 76, 320
 special effects, 79–80
 3-D, 77, 78
 2-D, 75–76
 on World Wide Web, 232–233
Animation clips, playing, 331–334
Animator, 331
 multimedia project team and, 173
Antialiasing, 60, 61
Appearance, 140
Apple Computer, Inc., 29, 81, 87, 186, 249
Apple Macintosh, 5, 29, 31, 37, 38, 51, 54, 59, 60, 64, 96, 125, 148, 186, 223, 262
Apple System 7.x, 125

Applets, 228, 229
Approach, multimedia titles and, 121–122
Archive Films, 247
Arial, 51
Arrange menu, 263, 301, 302
ASCAP, 246
Asymetrix, 96, 109, 133
Asymmetrical balance, 142–143
AT&T, 163
Audience
 interactive design and, 147–148
 target, 120, 200
Audio card, 35
Audio Video Interleave (AVI), 87
Authoring programs, 53, 94–109
 icon-based, 101–102
 time-based, 102–103
Authoring tools, Web-based, 109
Autodesk, 331
AVI. *See* Audio Video Interleave

Background, 269
Baker and Taylor, 206–207
Balance
 asymmetrical, 142–143
 screen design and, 140–144
 symmetrical, 142–143
Bandwidth, 250
Barnes & Noble, 208
L.L. Bean, 14
Beta testing, 134
Better Homes & Gardens, 211

Binary digits (bits), 33
Binary system, 33
Bitmap, 55
 jagged edges, 61
Bitmap graphics, 55–56
Bit, 33
Blockbuster, 208, 210
BMI, 246
BMP, 59
Boeing, 4, 15, 81, 163
Bold, 50
Bookmark, 151
Book metaphor, 98
Booktronics, 210
Borders, 208
Brøderbund, 165, 202, 211
Browser, 231–232
Budget, 169–171
Bundling, 209
BZ/Rights and Permissions Inc., 246

C++, 104, 228
Cable modem, 224
Card stack metaphor, 98
Catalog sales, 209–210
CD. *See* Compact disc
CD Audio, 185
CD-DA, 185
CD-E, 251
CD-I, 186
CD-R, 187
CD recorder, 40
CD-ROM, 35, 186
 distributing multimedia titles, 200–212
 labeling and packaging, 192–194
 mastering and replication, 189–192

suggested retail price (SRP) The price that the manufacturer specifies through labeling or advertising as the price to be charged by a retail store.

symmetrical A method for achieving balance in screen design through the arrangement of similar elements.

tags Sets of characters used in HTML to format the appearance of a document, such as centering a heading. For example, the tags <C> and </C> are used to center the words that appear between them.

target audience Those people determined by the developer to be the primary users and/or buyers of a multimedia title.

task analysis A study of the specifications for a multimedia title that results in a list of tasks necessary to create the desired title.

telecommunications A communications process involving phone lines, cable, or wireless transmission.

template A precise layout indicating where various elements of a multimedia title will appear on the screen. Templates can aid in the design process by providing consistency, shortening the development time, and preventing "object shift."

testing An ongoing process in multimedia development to help ensure that a product is meeting the desired objectives and specifications.

text A multimedia element consisting of alphanumeric characters.

time-based program An authoring tool that uses a movie metaphor in which the multimedia title is played frame-by-frame, and the screen is a stage where objects (text, graphics, and so forth) are displayed.

topical navigation A navigation scheme that allows the user to select from an array of choices or even search for specific information. Examples are multimedia encyclopedias, interactive shopping catalogs, and informational kiosks.

trademark A name, symbol, or other device identifying a product; it is officially registered with the U.S. government, and its use is legally restricted to its owner or manufacturer. Trademark protection covers the title of a publishable work and, in the case of fiction, often the name of its characters.

transactional kiosk A computer-based system that allows transactions, such as airline self-ticketing systems or bank automatic teller machines (ATMs).

tweening A technique in path animation in which an object's beginning position in one frame and ending position in another frame are set and the program automatically fills in the intervening frames.

type style An font attribute, such as bold, italic, or underlining, applied to text for emphasis.

unity In multimedia development, unity refers to how the various screen elements relate—how they "fit in."

upgrade kit A set of hardware (CD-ROM drive, audio card, speakers) and software that can be installed in a computer to create a multimedia personal computer.

user interface The way in which the user interacts with a multimedia title. The interface involves designing the appearance (how each object is arranged on the screen) and the interactivity (how the user navigates through the title).

value-added The concept of adding value to a CD title as it moves through the distribution process. For example, wholesalers can stock large quantities of a product, which is a value to the retailer, and retailers can provide convenient locations, which is a value to the customer.

video card A hardware device installed in a computer that allows digital images to be displayed on a monitor. The type of video card determines the screen resolution and speed at which graphics and text are displayed.

virtual reality (VR) A computer-generated environment that surrounds the user so that he or she becomes part of the experience.

Virtual Reality Modeling Language (VRML) The computer language used to create 3-D environments on the Web that allow a user to move through a space or explore an object.

volume The height of each peak in a sound wave which determines loudness; the higher the peak, the louder the sound.

VR See *virtual reality*.

VRML See *Virtual Reality Modeling Language*.

warping The process of distorting an image to create a special animation effect.

Web page A document that is written in HTML and forms the basis for the World Wide Web.

Web site A location on a server that is made up of Web pages that are linked through a navigation scheme created by the site developer (a person, company, or organization). In all sites there is a primary page often called the home page. This is generally the anchor that is used as a reference point by the site developer. All other pages link to this page and usually there is a navigation button that takes the user to the home page from the other pages.

wholesaler A middleman in the distribution process between the CD publisher and the retail store. The wholesaler often stocks large quantities of a number of titles.

World Wide Web (WWW) The part of the Internet that allows delivery of multimedia elements, such as graphics and sound, and for hyperlinking of content.

WWW See *World Wide Web*.

Yellow Book The CD-ROM (compact disc–read-only memory) format specifications. *Yellow* refers to the color of the binding of the document in which the specifications were first published.

© 1995, Landoll, Inc.
Ashland, Ohio 44805
® The Landoll Apple Logo is a trademark owned by Landoll, Inc.
and is registered with the U. S. Patent and Trademark Office.
No part of this book may be reproduced or copied.
All rights reserved. Manufactured in the U.S.A.

DAVY CROCKETT
STORYBOOK

Davy Crockett, dressed in shirt and trousers made from deerskin, and wearing Indian moccasins on his feet, was creeping silently through the Tennessee forest one December day.

Davy, who lived with his wife and children in a log cabin he had built from trees cut and trimmed with his ax, had a big problem. He depended on his rifle to hunt food for his family, and he had run out of gunpowder the night before.

He was one of the first brave pioneers to settle this far west, and there were no stores to buy supplies, and no neighbors from whom he could borrow.

The nearest cabin was many miles away. Winter was near, and a cold rain had already soaked him to the skin. Water had leaked in through his moccasins, and his leather shirt felt icy.

The only warm place he could find was under his hat, which his wife had made from a piece of coonskin with the fur still on it.

On his belt, Davy was carrying a horn from a woods buffalo which he had carved into a waterproof container for a day's supply of gunpowder. It was hollow, and he had made a wood plug from a walnut branch to fit in the big end and keep his powder dry. Only it was empty today!

He still had some bullets in the leather pouch he wore on the other side of his belt. He had also made these himself by melting lead over flames in the fireplace and pouring it into a mold. But they were no good without gunpowder.

A hunting knife in a sheath, and his rifle were the only other things Davy carried. The gun was so long that when he rested it on the ground it was taller than he was.

Settlers in those days had to remember a lot of things when they went out in the woods to hunt! Before Davy could shoot his gun, he had to pour just the right amount of gunpowder down the barrel. Then push a tiny patch of cloth, which his wife had cut from scraps, on top of the powder using a long wood stick called a ramrod.

Next, he pushed in a bullet, which looked a lot like a round ball, down the barrel on top of the patch, and another cloth patch on top of the bullet – otherwise it might roll back out if he pointed his rifle barrel down too far!

But he still wasn't ready to shoot. Davy had to sprinkle a tiny bit more gunpowder into a tiny pan on top of his rifle, where it would explode and set the powder in the barrel off when he pulled the trigger.

But knowing all these things didn't help Davy today. He was out of powder, and he must walk many miles in the cold rain, through the wet woods and rising streams, to get some more. His family needed food for the winter. And even though Davy was considered one of the bravest men on the American frontier, he had a lot of things to worry about today.

He had to keep a close watch in every direction, for there were wild animals in the forest and Indians that sometimes attacked settlers who had taken land the tribesmen felt belonged to them.

He walked up steep hills and through deep valleys. When he came to a small river that he had always waded across, the water was so high that he had to hold his gun high over his head to keep it out of the icy flood.

When he reached the other side, he was soaked to the skin and very cold! He ran to get warm and, toward night, finally saw the dim light from the cabin he sought.

He spent the night steaming in front of the fireplace, and the next morning, after his friends had fed him enough for two or three men, he was ready to start home.

By now there was a skin of ice over the cold water, but it wasn't thick enough to hold him. And when he got to the river this time, he had to cross it twice!

His gunpowder came in a keg, which looked like a small barrel and was heavy. Even though Davy was one of the strongest men around, he couldn't hold the keg and his gun over his head at the same time while pushing through the cold, icy water.

He tucked his gun under a bush on the river bank, and waded across holding the keg out of the water. Then he had to go back and get his gun.

Everybody at home was glad to see him when he finally arrived. And after Davy had a good rest that night, he was able to go out in the woods and get enough food to feed his wife and children until spring when they could plant a garden and have corn and potatoes, and onions, and lettuce again.

Soon berries would be ripe, and that summer they would get a cow, and they could have milk as well as meat and vegetables on their table.

Even after he became famous for many other brave and adventurous things he did, his children always told the story of the trip Davy had taken that winter so he would have food for his family.